Building Decisions

Decision-making is critical. Leaders make decisions daily, often hyper-focusing on cost at the expense of value for society and with negative impacts on the environment and the climate. Project teams are poorly equipped to make decisions and prefer to avoid conflicts instead of having a healthy discussion based on different perspectives. We need to learn ways of making decisions that give better outcomes.

Building Decisions presents a framework and practical guidelines for everyday decision-making. The authors introduce the reader to a specific decision-making system, Choosing By Advantages (CBA), and explain the CBA Tabular decision-making method. Providing detailed explanations using relatable examples such as choosing a phone and renovating a kitchen, the book also uses case studies of large construction projects from the authors' professional careers, including a museum in San Francisco, a tunnel in England, and the main railway station in Munich, to explain the CBA method. These varied examples will help the reader to improve their own decision-making process through critical reflection.

Written by three curious experts who founded a research lab focusing on decision-making and want the world to make better decisions. Paz, Annett, and Randi have published dozens of peer-reviewed papers and trained multiple practitioners and engineers on how to make collaborative decisions using CBA. They have designed this book to help professionals, leaders, anyone, and everyone make better and more inclusive decisions and achieve better outcomes.

Paz Arroyo is a global leader in Lean Construction. She earned her PhD at UC Berkeley, focusing on collaborative decision-making. She has tested the effectiveness of Choosing By Advantages (CBA) in construction projects, publishing widely on the topic. As a Lean and Quality leader for a major international contractor, she develops strategies to improve the construction industry.

Annett Schöttle earned her PhD at Karlsruhe Institute for Technology (KIT), Germany, on requirements and incentives to create a collaborative Lean project delivery system for capital projects. She works as an independent Lean consultant with extensive experience in megaprojects, helping teams collaborate to develop, align, and optimize production planning from project start to commissioning, and implementing CBA for better decision-making in various settings.

Randi Christensen is an engineer by background and earned her PhD at Aalborg University, Denmark, focusing on how we learn from project to project. As Sustainability Director at an international engineering consultancy, her focus is on decision-making as this is the key point where we can drive the projects in a more sustainable direction by including multiple voices and pools of knowledge.

Building Decisions

How Choosing by Advantages Drives Project Success

Paz Arroyo, Annett Schöttle, and Randi Christensen

Routledge
Taylor & Francis Group
LONDON AND NEW YORK

First published 2025
by Routledge
4 Park Square, Milton Park, Abingdon, Oxon OX14 4RN

and by Routledge
605 Third Avenue, New York, NY 10158

Routledge is an imprint of the Taylor & Francis Group, an informa business

British Library Cataloguing-in-Publication Data
A catalogue record for this book is available from the British Library

ISBN: 978-1-032-85466-3 (hbk)
ISBN: 978-1-032-84776-4 (pbk)
ISBN: 978-1-003-51831-0 (ebk)

DOI: 10.1201/9781003518310

Typeset in Times New Roman MT Std
by Newgen Publishing UK

Contents

Foreword

This is a very important book. We all make decisions and want to make good ones, but there are different ways to make decisions, and some decisions are better than others. This book provides compelling arguments that Choosing by Advantages (CBA) is the process you need—a process that is collaborative and inclusive.

The authors' credentials are impeccable, both as regards academic degrees and practical experience. As it happens, their degrees and experience are from the construction industry, but what they teach about decision-making can be applied in any industry and even in daily life. They argue persuasively, from reason and evidence, that Choosing By Advantages is the best way to make decisions; best because it promotes collaboration, provides a rationale for why a specific decision was made, helps build consensus, and enables continuous learning. There's more! Choosing by Advantages speeds up making decisions and making decisions that stick, even those involving many stakeholders dealing with difficult choices.

Jim Suhr created Choosing by Advantages while working with the U.S. Forest Service and published *The Choosing by Advantages Decisionmaking System* in 1999. In this book, the authors build on Suhr's ideas about choosing by advantages. They present research to prove its practical use in the construction industry and measure its benefits compared to traditional methods, such as Weighting Rating and Calculating (WRC) and Analytic Hierarchy Process (AHP) (see Chapter 7). They also connect CBA with our current knowledge of biases and highlight the relevance of being inclusive in the teams that make decisions, which is highly relevant for the construction industry and all other types of projects. They also question traditional decision-making practices in the industry, including the hyper-focus on first cost and the lack of understanding of value. In addition, they take a deep dive into traditional procurement methods and the use of WRC in the bidding process (Chapter 8). Finally, and by no means least, they discuss the implications of decision-making for sustainability (Chapter 10).

By the time you finish Chapter 10, you will very likely be thinking how to put what you've learned into practice. That's where Chapter 11 comes into play, providing guidance and encouragement to make a difference by speaking up for and using Choosing by Advantages.

Do you now make decisions by weighting the relevant factors or by treating cost as a factor? You will learn that these are two common mistakes in decision-making. Weighting factors is pre-determining their relative importance. This may seem reasonable, but different stakeholders in the decision-making process will likely have different ideas about what is most important to them. If their support is needed to act on a decision, advance rejection of their preferences is not a good place to start. If you want to have everyone involved agree on the right decision to make, first look at what alternatives are available and evaluate each of them against factors relevant to making the decision. Choosing by Advantages defers evaluation until the attributes of each alternative are agreed; first description, then evaluation.

If you want to select the alternative that offers the best value for money, don't use cost as a factor. Too often, when cost is considered a factor alongside others such as "ease of implementation," cost is treated as the factor with the greatest weight. This results in discarding the more costly alternatives without considering the advantages they offer. CBA defers consideration of cost until there is agreement about the total importance of advantages offered by each alternative—a prerequisite for making decisions based on value for money. Solution: do not treat cost as a factor, but rather as the price to pay to get the advantages offered by an alternative. If you discard alternatives without first considering their advantages, you risk spending more and getting less.

Does your current way of making decisions help create transparency so everyone can understand why the decision was made? Does it help build consensus so people agree on what is important and are ready to help put the decision into practice? Does it result in best value for money decisions? If not, read this book.

Glenn Ballard

Preface

Back in 2018, it was clear to us that decision-making in the construction industry needed improvement. That is why we decided to found and finance a research lab dedicated to doing that: CollabDecisions.com. We thought of writing a book together for a long time, but our busy schedule, full-time jobs, and being in different countries and continents made it difficult. It was not until July 2023, in Paris, France, that we made this a priority. We stayed in a week-long retreat together in a beautiful apartment we rented in the suburbs, and we covered it with sticky poster papers and sticky notes. We had great discussions and productive iterations but finally had a game plan! And we were excited to share it with the world.

This book is the culmination of our collective years of collaboration and learning, dedicated to helping organizations and project teams make sound decisions focusing on inclusion and sustainability. Over the years, we have published numerous peer-reviewed papers, trained multiple practitioners and engineers on collaborative decision-making, and consulted for large firms and projects, leaving a tangible impact on the organizations we work with.

Who we are

Paz Arroyo earned her PhD at the University of California, Berkeley, where she started asking how design teams made decisions about their buildings, and how they included sustainability considerations. She researched and compared different decision-making methods to support project teams. She found strong evidence to support that Choosing By Advantages (CBA) is the most appropriate method to help project teams in the design and construction phases, and we will explain this in detail throughout this book. Paz was an Assistant Professor at the Catholic University of Chile and a Lean Consultant in the USA. She has published multiple articles on CBA implementation and its benefits and taught CBA internationally. Paz works at DPR Construction as a Quality Leader, a forward-thinking general contractor where CBA is considered a best practice for making complex project decisions; she continues to help teams make decisions, learn, and write about what they have learned in collaboration with Annett and Randi.

Annett Schöttle did her PhD at Karlsruhe Institute for Technology (KIT), Germany, on requirements and incentives, based on the self-determination theory, to create a collaborative Lean project delivery system for capital projects. Alongside her work as a research associate, she gained industry experience as a consultant in claim management. Annett works as an independent Lean consultant, with extensive experience in megaprojects, helping project teams collaborate to develop, align, and optimize production planning from project start to commissioning, and implementing CBA for better decision-making in various settings. Her research and practical insights have been published in academic journals, books, and conference papers. As a trainer for Choosing By Advantages (CBA), she has developed a CBA implementation concept for a German automobile manufacturer, and educated facilitators worldwide. She also facilitated the CBA workshops for the International Group for Lean Construction and for the German Lean Construction Institute.

Randi Christensen is an engineer by background and earned her PhD in Construction Management, Learning, Lean, and Innovation. She was head of Lean Construction DK for six years, and now works as a Sustainability Director at COWI, an international engineering consultancy, and as a board member at Aalborg University, one of the world's highest-ranking engineering universities. She keeps focusing on decision-making as a means to change the construction industry and drive the projects and organization in a more sustainable direction by including multiple voices and pools of knowledge. By training others in CBA and introducing the method to projects, she has proven that decision-making becomes more efficient and brings more innovative and sustainable solutions.

Even though our experiences have mainly revolved around the world of projects in the construction industry, we have also consulted for politicians choosing which new project to build in their city, owners in the automotive industry deciding on the best project delivery system for their new manufacturing plant, and the waste management industry choosing how to design an efficient residential waste management collection system at a city level. The lessons learned that this book presents apply to any group that wants to create a resolute action and implement a decision based on shared understanding and transparent trade-offs. Which, from our experience, is almost any group of humans with the task of creating something new.

The structure of the book

In Chapter 1 "What is a decision?," we start by presenting the elements of a decision to help you identify when you are making one. We examine a bit of the history to understand our current decision-making practice and conclude that we should manage decision-making as a process and involve multiple perspectives.

In Chapter 2, "Why decisions do not stick," we discuss why decision-making is often frustrating and what it takes to make a decision that sticks. People make decisions, and we need to understand that people do not always act rationally. We will show the importance of a transparent decision-making process and the necessity of understanding the purpose of a decision. We will discuss negative iterations and explain the advantages of making decisions at the last responsible moment. Furthermore, we talk about considering emotions and biases impacting a decision. We will discuss creating a psychologically safe room for people to speak up and including people with different perspectives to increase the chances of making a sound decision that different stakeholders support.

Chapter 3, "Focus on value and cost," discusses the assumptions we make when we hyper-focus on cost. We cannot just focus on cost; decisions that stick over time call for incorporating multiple perspectives, such as quality, productivity, safety, and environmental and social impacts. We need to focus on both value and cost.

Chapter 4, "Changing bad habits," will show bad habits when making decisions and the consequences of doing so. We present essential ideas from the Choosing by Advantages (CBA) system to make sound decisions instead. We will first discuss the relevance of learning a new method and remind you that it takes time, patience, and trust. We make these bad habits explicit and show you how CBA helps us avoid them. Furthermore, we will introduce CBA vocabulary and ideas.

Chapter 5, "CBA: a sound decision-making method," describes the Choosing by Advantages System. We build this by showing a step-by-step example of a simple decision people can relate to, in this case, choosing which phone to buy. Through examples, we present a new vocabulary and principles to apply practically. This simple decision is written from a single decision-maker perspective.

Chapter 6, "The CBA Tabular Method: Sustainable Museum Case Study," presents a case study based on research on a very complex multi-million-dollar decision involving multiple stakeholders. This example is still simple to understand and explains how a team designing and building a new museum can use the same simple steps and see the benefits of this approach in practice.

Chapter 7, "Not all decision-making methods are created equal," presents research studies and practical applications to understand better the benefits of using CBA and learn more about the implementation process. Every case will focus on a different aspect of CBA, such as focusing on relevant differences between alternatives and understanding the experience of the decision-makers and the results of the decisions. Furthermore, the cases will help to better understand CBA compared to traditional decision-making methods. This information is primarily relevant if your organization already uses traditional decision-making methods.

Chapter 8, "CBA and the tendering procedure of construction projects," will examine the tendering procedure and its impact on projects. We will analyze the current decision-making approach and mindset and provide an outlook on the necessary changes to apply CBA. Through two case studies, we will further illustrate the incorporation of two central decisions: the decision of the project delivery system and the evaluation of proposals. After that, we will present how you can apply CBA in the tendering procedure based on a simplified example.

Chapter 9, "Building a collaborative decision-making process," offers strategies and examples of key elements of a governance process, from preparation to documentation of a decision. We also introduce Lean ideas that support smart and efficient decision-making. Lean is a philosophy and a set of management methods and tools that aim to increase value for the client while reducing waste. With a collaborative decision-making process that includes a more comprehensive view, you are more likely to stay true to the CBA principles, even if you need to adapt them along the way.

Chapter 10, "Decisions and innovations for our future," will provide an outlook on how to cope with current and future challenges. We know we need to change to deliver the services needed for our communities while reducing its harmful impacts on climate, nature, and people. This will increase the complexity of project management, where we must balance many contradicting interests. CBA can be a catalyst that drives innovation to create more sustainable solutions.

Chapter 11, "How to build better decisions?" is the conclusion. We will end the book with a summary and give guidance on how to make better decisions by including multiple perspectives. We will leave the reader with some tangible steps to follow and questions to expand on the reflections on decision-making methods. We will also invite more people to elaborate, implement, and research the topic for the benefit of all.

Who is this book for?

This book could be relevant for many people inside and outside the construction industry. We have written the book in a sequence that builds up the capacity you need to support better decision-making, and we recommend that you read it all. However, depending on your role, you might dive deeper into some of the chapters.

You might be an owner, an investor in the construction industry, or a politician making strategic decisions for your community. You might have a clear vision of a project but find it difficult to communicate your vision to a team of specialists from the industry. This book will help you ask the right questions to the team, which will help you make better decisions. If you understand what a decision is (Chapter 1), you should ask your team to present the alternative solutions where you can assess both value and cost clearly and in a structured way so you can decide (Chapters 3 and 5). Also, you might consider using CBA for tendering (Chapter 8) or building some requirements for collaboration in a governance process (Chapter 9).

You might be an architect, engineer, contractor, or civil servant with sector-specific knowledge and insights. You base many decisions on your professional expertise and experience and find it frustrating that clients, managers, politicians, or collaborators keep challenging your professional knowledge. This book will help you understand why current practices might make it difficult to make decisions stick (Chapter 2) and which habits we need to change to deal with this (Chapter 4). Through a detailed case study, we will show how CBA enables professional knowledge to be valued in decision-making (Chapter 6), strengthens arguments, helping the client to decide, and how you can help the industry deliver more efficient and sustainable solutions (Chapters 9 and 10).

You might be a student or new to the industry and still trying to get a foot in the door. You might often ask yourself if you should speak up and suggest other alternatives, but find this difficult as people around you are much more experienced. By understanding the basics of decision-making (Chapter 1) and how CBA (Chapter 5) and other multicriteria analysis work (Chapter 7), you can challenge the current practice and build your career to create a more sustainable world (Chapter 10).

Whoever you are, we hope you will find inspiration and solutions. Through decision-making, we can accelerate change and introduce innovations needed to improve current practices. The decisions you make today will shape the future, and with the challenges we face regarding climate change, biodiversity, and social injustice, we need you to contribute to the solution. The construction industry will continue to serve the fundamental needs of people worldwide, such as housing, education, transportation, healthcare, knowledge advancement, and manufacturing. We want you and your project to have access to the best knowledge about decision-making methods and processes to ensure you get the best value for the resources you have available.

Acknowledgments

We are truly grateful for all the support and feedback we received while gathering cases, researching, and writing this book. This book celebrates the value of collaboration and sharing of success. We extend our gratitude to everyone who has discussed, challenged, and built on our expertise and experiences over the years. It has been a team effort, and we would like to call out some of those who stepped in when needed.

First, we thank Cowi-Kampsax-Personalefond for funding our writing camp in Paris in July 2023. This funding allowed us to meet in person, kick-start compiling cases and ideas, and develop a strategy for our writing process.

Cases and collaborators

Throughout the book, we reference theses and papers based on a number of cases, and some deserve special thanks. Thanks for believing in us, trusting us to support some of your most important decisions, and taking the time to help us learn and document the results. The experiences from the practical application of theories and methods make this book stand out, and we hope you all will continue to support the practical application of research and develop new knowledge.

Thank you to Aalborg Havn and Mette Schmidt, Chief Technical and Sustainability Officer at Port of Aalborg, for experimenting with collaborative ways to make projects more sustainable and investing in thorough Lessons Learned to benefit the wider industry.

Thank you to Metroselskabet and Russell Saltmarsh, Project Director at Metroselskabet & Hovedstadens Letbane, for investing in innovation projects to create new solutions for a more sustainable metro line. This inspired the use of CBA to create innovative solutions.

Thank you to Copenhagen Municipality, the Department of Technical Services and Environment, and Manager Henriette Hall-Andersen for taking the initiative to train facilitators and implement new and better decision processes.

Thank you to the University of California San Francisco and their former campus architect and associate vice chancellor of Capital Programs Michael

Bade, Principal Campus Counsel Christine Haas Georgiev, and Vice President Health Major Capital Projects Stuart Eckblad for their innovative work in improving the project delivery system in the public sector and for their out-of-the-box thinking. Thanks for the opportunity to collect data from UCSF cases.

Thank you to Deutsche Bahn Infra GO, former Head of Engineering at Munich Main Station André Haubrich, and the entire project team for embracing new and innovative approaches to understand and manage the project according to its specific needs. Thanks for your trust and the opportunity to apply Choosing by Advantages.

Thank you to the Exploratorium Museum in San Francisco, California, its former Relocation Project Director, Kristina Woolsey, and its current Project Director of Campus Facilities, Shani Krevsky, and the operation team for being open to using CBA and for their leadership in sustainable building design and operations.

Thank you to National Highways, and particularly the Lean team, for investing in and supporting the application of Lean Construction on major projects. This gave the opportunity to train facilitators and apply new practices for collaborative decision making.

Thank you to our COWI and DPR Construction colleagues and collaborators who have supported the implementation of Choosing by Advantages on numerous projects. You have seen the method's benefits and found new ways to use them for the benefit of clients and society.

Reviewers

We want to thank friends, colleagues, and family members who reviewed chapters or even the whole book to ensure the content was clear and understandable. Thank you to Thomas N. Christensen, Ricardo Henoch, Glenn Ballard, Louise Vinther Alis, Charlotte Bruun, Michael Bade, Christine Haas Georgiev, Morten Skaarup Jensen, Tommy Olsen, Alistair Kean, Vic Ortiz, Johannes Pitterle, Ron Migliori, Cheryl Parsons, and Hao Tran.

Personal notes

My profound thanks to my husband, Ricardo, for all his support and love over the years. I would not have succeeded without you by my side. Thank you to my daughter Clara and son Ricardo for their patience and for cheering me on so I could finish writing. You both inspire and bring me joy. Thank you to my mother, Carmen, for supporting us in our journeys far from home. (Paz Arroyo)

Throughout my journey, I have encountered many individuals who inspired me, offered opportunities for learning, and contributed to my growth both professionally and personally. I am deeply grateful to all of these people, but I want to extend a special thank you to my parents, Gisela and Günter Schöttle, for their unwavering support and belief in me. (Annett Schöttle)

I would like to thank close family and friends for helping out and understanding when workload, book writing, and personal events collided. Particularly, I would like to thank my children, Isabella, Alexander, Emilie, and my husband, Thomas. Without you, this would not have been possible. You are my love, my inspiration, and my motivation. (Randi Muff Christensen)

Introduction

We are *building decisions*, literally and figuratively, in our lives and the projects we work on. With this book, we want to raise awareness of how we do that and how the methods and processes we use to make decisions significantly impact our decision outcomes.

Construction projects involve many stakeholders, such as authorities, designers, contractors, operators, users and the community. Every project has a beginning and end, a limited budget, and a team of people joined to work together to build something that was not there before. All those people must make multiple significant decisions regarding the functions, methods, and design to ensure the project fulfills the client's needs. Everyone involved has their own experiences, views, and interests, which creates a challenging set-up for making decisions as a team. Ensuring stakeholders consider decisions and listen to each other, even under time or budget pressure, requires work. Having a shared process and method for decision-making will help.

The challenges mentioned for construction projects are not unique to the construction industry; software developers and film producers operate similarly. Many organizations also work on activities that are treated as projects, such as developing a new training program or a new customer service process. Moreover, in our personal lives, we manage projects, such as going on vacations, planning a party, or renovating our homes, where we need to make decisions as a team, with friends, or with family. Therefore, our insights from years of professional experience in the construction industry are relevant in other domains where people must collaborate to deliver a product, an experience, or a process and make decisions to shape that new thing.

One of the issues that made us particularly curious about how we make decision-making in the construction industry was that we often see decision-makers, leaders, and managers making decisions based solely on cost. Low cost is seen as the ultimate goal, and often, the time spent talking about cost in meetings significantly outweighs the time spent discussing the value you wish to provide to the project stakeholders. We also see this in public procurement, where cost is typically the factor given the most weight when choosing which construction company wins a project. The industry has an underlying

DOI: 10.1201/9781003518310-1

assumption that we are rational, logical, thinking creatures and that decision-making is a simple equation to bring down costs. But still, psychologists point out that this is not the case. Emotions, values, and previous experiences influence our decisions.

We need a better way of making decisions to build a better future for us all. We need to design our decision-making practice to nudge us to talk and listen to different opinions and allow project team members to discover ways to increase value for stakeholders in the long term while using our resources efficiently. This book invites you to learn about a sound decision-making method and collaborative decision-making process.

The three of us collectively bring over 50 years of professional experience to the table. We are continually astounded by the discovery that significant decisions for large construction projects are often made based on the gut feelings of selected individuals who did not collaborate with the team to understand the consequences and who hyper-focus on a cost estimation with unclear assumptions. We usually ask project teams about their decision-making method and process. More often than not, people cannot provide a clear answer. Most people are not aware of how they make decisions. In the construction industry, where many organizations and contracts are hierarchical, project teams tend to wait for others in power to make the decisions. As changemakers, we need to recognize this gap in understanding when we introduce the topic of decision-making. Given all the myriad challenges we face in social, climate, and environmental issues, we need to take decision-making far more seriously. We need a process that helps teams better understand different perspectives, align their interests, and create more transparent decisions that become easier to understand, document, and communicate. A collaborative decision-making process and a sound decision-making method allow the project team to move forward in the same direction with confidence and resolution. Unfortunately, very little has been written about how we can make decisions collaboratively while still being efficient. We are happy to provide insights based on our research and practice.

This book is the culmination of our collective years of collaboration and learning, dedicated to helping organizations and project teams make sound decisions focusing on inclusion and sustainability. We are Paz, Annett, and Randi, three PhD researchers and friends who founded and financed a research lab, CollabDecisions.com, in 2018. Over the years, we have published numerous peer-reviewed papers, trained multiple practitioners and engineers on collaborative decision-making, and consulted for large firms and projects, leaving a tangible impact on the organizations we work with.

1 What is a decision?

What was your last decision, and how did you make it? You might not notice, but we make decisions all the time. Some decisions are big, some are small, and this chapter will start by defining "a decision" and elaborate on why some decisions should get more attention. A shared understanding of what constitutes a decision enables a more thorough consideration before jumping to conclusions, and as some decisions have significant consequences, we ought to have a strategy for how to deal with these. Throughout history, we have used and developed decision strategies and practices, but with the challenges we face today, these strategies and practices need an update to allow for more cross-disciplinary collaboration. Therefore, we conclude this chapter by elaborating on why a decision is a process and what elements it should contain to include more perspectives.

When is a decision about to be made?

To improve the way we make decisions, we need to first recognize when we are about to make one, which can be rather difficult. Therefore, we will list some key elements of decision-making so that you can more easily recognize when you make decisions. *The first requirement for making a decision is that you have two or more alternatives to choose from*, see Figure 1.1. You could choose between different types of kitchen floors when renovating your house; you can choose between different types of software for your company, different types of materials for the structure of your building, or you can choose different types of cars or phones, depending on your needs, etc.

If you only have *one* valid alternative, no decision is needed, and we can move ahead without further consideration. However, it can be difficult to recognize if we have more than one alternative and thereby need to make a decision. If we have one valid alternative at hand, we could be tempted to move ahead without actively looking for more alternatives. This often happens when we feel under pressure, and our System 1 thinking takes over, as Kahneman (2011) describes. This means that we make a quick, often unconscious decision based on previous experiences or beliefs that may or may not be relevant

DOI: 10.1201/9781003518310-2

Figure 1.1 A decision requires two or more alternatives

to the current situation. We do not stop and reflect but jump ahead and act automatically.

In projects, we are often paid to deliver *a* valid solution: a bridge design or a description of a construction process. If we need to move fast due to time and budget constraints, we might not be looking actively for alternatives. This could be due to the assumption that looking for alternatives and then needing to assess these could add complexity and take additional time and resources. So, we might prefer, consciously or unconsciously, not to look for those alternatives. In other cases, you might believe someone else, perhaps someone with more authority than you, has already decided on the path forward, so you do not look for alternatives or consider sharing other ideas you may have. Again, an automatic reaction makes you act instead of stopping, reflecting, and communicating your thoughts.

But just because you are unaware or do not recognize the alternatives doesn't mean they are not there. Moving ahead without considering an alternative or speaking up about your ideas is also a decision you make, even though you do it in a split second, and this might exclude you from clearly considering alternatives and the benefits these might bring. It also excludes you from the opportunity to learn from your decisions, as you have not recognized that you made one. Try to imagine alternatives for a decision you are to make, just to practice looking for them. When you have more than one alternative, you now know you are about to make a decision. You can choose to do one thing or choose something else—but you can't do both.

So why all this talk about alternatives? If you have a valid alternative, wouldn't it be easier and cheaper to move forward with this? Why not just use the well-known methods and designs? Well, yes, it might have been easier, and that's probably why we have System 1 thinking to support our immediate decision-making process. However, we are facing significant challenges nowadays, for example, the climate or the biodiversity crisis, and these challenges require us to think of better and more sustainable alternatives, no matter which industry we work in. For example, we have used the same concrete mix for decades, with cement as a central ingredient. These are well-tested processes, and the risk of defects is minimal, but at the same time, the cement industry emits 7% of global carbon emissions. When we know of more sustainable alternatives to the traditional mix, should we not consider these? They might be new and they have been thoroughly tested and we have durability studies available. Do we dare to change? Do we dare not to? Many more such decisions will be required to change our industry and society, and we need to be able to make these decisions based on the newest knowledge and with consideration to the changes of political agendas.

Some might also think that doing nothing is not a decision. We are prone to the status quo, staying where we are, using the techniques we know. The well-known path is comfortable and easy. However, if there were an alternative, including the use of a new technology, then doing nothing or just following the practice we are used to in the situation, is also a decision with consequences. You had alternatives but stayed with the default and might have missed out on benefits or opportunities. However, these decisions of doing nothing will not be documented or communicated, so you miss the opportunity to learn from the decision. Also, the process excludes others from contributing to the considerations. We often forget these decisions when we do nothing and no actions are taken to change the status quo.

We are programmed to make decisions automatically, and therefore, we must train ourselves and our teams and collaborators: (1) to look for more alternatives; and (2) to make decisions more intentionally. It can be difficult to recognize that you are in front of a decision and need to take a stand, particularly if you are not looking for other alternatives. So, looking for a "Plan B" (and Plan C and Plan D, etc.) might be helpful.

Always consider Plan B

Randi was working on a major infrastructure project consisting of a highway tunnel and connecting roads. In a very early phase of the project, while selecting different options for the river crossing, a sketch of the tunnel was made. The tunnel length was sketched to be the shortest possible to indicate a potential and valid option. However, with new knowledge about the strategy and values of the client, the tunnel team thought the project would deliver more benefits if the tunnel was lengthened.

The immediate response from the client's representatives was that they were concerned that a longer tunnel would mean a prolonged schedule, which was perceived as the main constraint for the project. The team had a valid option at hand, so the client did not wish to go ahead and spend more time considering a change. However, the project team decided to assess a potential alternative for a longer tunnel and initiated a decision workshop that included more disciplines and perspectives to identify and record the benefits of both the minimum tunnel length and a longer tunnel.

Not much design work was done in preparation for this, and the principal differences between the minimum length and the longer tunnel alternatives were assessed on a general level before diving into the details. The longer tunnel would allow the road to move further away from environmentally protected areas and local communities before it surfaced to ground level. The process of boring the tunnel was at the schedule's critical path; therefore, a shorter tunnel had an advantage here as it was quicker to construct. However, adding up all of the advantages for each of the alternatives, the longer tunnel had the most important advantages. Furthermore, no significant difference in cost was identified. When the client was presented with the alternatives and the benefits, a decision was made to extend the tunnel.

The case showed the importance of always being open to considering more alternatives, despite having one valid alternative at hand (Christensen et al., 2022). The case also showed that constraints we perceive as cast in stone sometimes can be negotiated. We might reject other alternatives based on our assumptions of minimum requirements and constraints. But as the public agenda changes to favor more sustainable and environmentally friendly alternatives, we might need new strategies to truly consider all alternatives.

This gives us a responsibility to look for more alternatives and recognize when facing a decision. However, we should not always dream up all kinds of unrealistic alternatives that fail to meet even the most basic needs and requirements. But be curious as to why we might immediately reject some alternatives—there might be a hidden gem here in the form of a new alternative or innovative solution. Being aware that a decision is to be made enables you to plan for the decision-making process and deliberately consider the alternative by using System 2 thinking, i.e., slow thinking (Kahneman, 2011). This also allows you to engage a group of diverse stakeholders to contribute. But to do so, we will look further into what constitutes a decision and how it is made.

When has a decision been made?

A decision is a "Choice of action – of what to do or not to do ... made to achieve goals" (Baron, 2008). This is a crucial understanding of decision-making; as time moves forward, your actions have consequences that could be irreversible, as resources will be allocated, and it will be impossible or extremely costly to change back to the situation that existed before (Howard, 1966). A decision will always be followed by actions, and the actions will lead to consequences. When you decide, you must therefore define what actions will follow and what consequences to expect from these, see Figure 1.2. You and your team must act

Figure 1.2 A decision leads to actions and consequences

according to the decision and consider whether this will require changes to the design, plan, or budget, or if there will be any other actions needed to implement a certain decision.

In decision-making, you try, perhaps without noticing, to predict your decision's consequences. What will happen in the future if you choose a long or a shorter tunnel? You try to foresee the consequences and compare these for different alternatives. This requires imagination, input of data, and experiences from past decisions. This also means that decision-making isn't an exact science but an imperfect prediction of the future with all the inherent uncertainty. This brings a dilemma: decision-making is a prediction of what will happen, and at the same time, decisions are irreversible with consequences. Decision-making is, therefore, very much linked to time. Linking decision-making to time tells us a decision has a before and an after:

1. A "before" includes the realization that you have two or more known alternatives.
2. An "after" is when you act according to your decision.

At the "before" stage, you realize you have alternatives, gather intelligence, try to predict the consequences, and compare to find the best outcomes. At the "after" stage, you act according to the agreed actions and start to see the consequences. You can also assess whether the consequences correspond to what you predicted, giving you a golden opportunity to learn and refine your practice and expectations for future similar decisions. Decision-making, therefore, includes several actions or steps, and we argue that decision-making should be regarded as a process.

By making a decision, you commit to moving forward in a certain direction by taking related actions, which will lead to consequences. Making up your mind but not acting on it is not a decision. No, thinking about the diet or signing up for the gym is not enough! You also have to act on it.

How we make decisions matters

Have you ever thought about which method you use to make decisions? Of course, a good decision-making method will not guarantee that you will make a good decision, but it can increase the chances of getting it right. Many people consider decision-making to be a point in time (Garvin and Roberto, 2013). A single action or a thought in the head of the decision-maker. However, we will argue that *a decision is not a magic point in time but a process involving communication and negotiation between individuals or groups*. Furthermore, most decision-makers focus on the content and not the process (Kahneman et al., 2013), and therefore, they do not realize they are jumping ahead to actions while missing an important step: choosing the right method to support decision-making.

The following quote, by Jim Suhr (1999), summarizes the importance of choosing a good decision-making method:

> Among the most important of all the decisions the world's people will ever make are their decisions about how to make decisions, because their decisions about how to make decisions will strongly influence all the other decisions they will ever make. Furthermore, human performance — including organizational performance — is a decision-making process. Therefore, by improving the way they make decisions, the world's people will be able to make substantial improvements in both individual and organizational performance. And this will improve their quality of life.

To ensure we are not jumping into actions, which would be our inbuilt preference (System 1 thinking), we can use different methods to gather input, reflect on preferences, and take the time to consider alternatives, see Figure 1.3. The decision-making method also has an impact on how to involve people to participate in decision-making.

An example of the importance of decision-making methods can be observed by considering the different types of democracies in the world. For example, in direct democracy, the people vote on laws and policies. This gives all citizens the right and the obligation to stay updated, have an opinion, and vote when a new law is needed. Direct democracy gives all the power to the voters and the citizens, and there is a risk that the tyranny of the majority will prevail, which might harm the minorities. Another type could be a representative democracy, where the people vote on those governing on their behalf for a defined period. Here, the citizens are not as engaged and only vote, for example, once every four years. Representative democracy has the risk that minority views are underrepresented. Here there is also a potential risk that minority views are underrepresented. In both types of democracies, there is the opportunity to develop more efficient processes for updating laws and policies, and it is possible to set up a system to protect a debate that includes minorities. This is a complex and often debated topic, since we know that the methods and processes to support democracy have an impact on important decisions.

In this book, we look at decisions that might not be straightforward and where several factors should be considered. Some of the complexity in decision-making stems from factors that are not directly comparable. One alternative might be preferable when considering one factor, while another alternative will be better when considering other factors. For example, would you rather take

Figure 1.3 Choosing the right methods for decision-making is important

the bike to work and get the exercise or take the car and be able to bring back groceries? Here, the benefits of exercising or bringing back groceries are not directly comparable, and you cannot objectively and universally measure the trade-off. When you scale this challenge to bigger decisions with two or more alternatives and several factors, it becomes obvious that a systematic approach to decision-making can be useful.

We are looking at methods that help us choose and create a shared understanding of why an alternative was chosen. We are also mainly considering decision-making that has consequences for more than one person or organization and where input from different sources is desired. Finally, by having processes and a shared vocabulary for decision-making, we distinguish between the people involved in the decision-making and the actual decision. By separating the people involved and the decision, those involved might find it easier to disagree and make constructive and more informed decisions (Garvin and Roberto, 2013).

Therefore, it is necessary to have a shared vocabulary to discuss decision-making and a process that allows you to plan for a decision. With this new perspective on decision-making and its components, we will jump back in time to understand who we should include in this decision-making process.

Who had the right to make decisions in the past?

We perceive who can and should be involved and finally make decisions based on our cultural history of how we organize our societies. Understanding where we come from as a society could help you understand why we act as we do today. Therefore, we will go a bit back in time to understand who the decision-makers were.

In ancient Greece, the plebs, i.e. the common people, loved Greek dramas. The dramas told a tale of helpless humans whose fates were in the hands of the Gods. If the Greeks wanted future predictions, they didn't turn to wise philosophers but to oracles (Bernstein, 1996). They trusted the Gods to point to the best choice. Also, knowledge of the past was shared through storytelling and myths, and accumulated decision-making wisdom was passed on but with little responsibility of the individual.

This changed with the emergence of democracy in occidental cultures. Now, every male citizen could participate in the main body, Ekklesia, a gathering of citizens of a city who met monthly to decide political issues. Every man could speak up and vote on decisions by raising their hand. They could list decisions related to all state matters, e.g., military questions, financial decisions, food supplies, etc. A smaller decision body, the boule, was selected by rotation. This group prioritized the topics discussed at the assembly, the Ekklesia. Even though all male citizens were allowed to speak at the assembly, the boule approved the agenda and did not have to consider all input (Rhodes, 2015). Also, in a crisis, they could decide without calling the Ekklesia. Through this system, some roles and responsibilities were defined in decision-making, and more people could have a say in an organized manner.

The different forms of government applied throughout history since then mainly differ in the application of decision-making: how many were considered full citizens and therefore could participate in the decision-making, and what decisions should be made by the authorities and which by the assembly (Rhodes, 2015). Noteworthy, however, it wasn't until 1893 that New Zealand became the first country to grant women voting rights; still today, women do not have voting rights in all countries.

Developing and understanding probabilistics in the Age of Enlightenment led to confidence in predicting the future (Bernstein, 1996). With the mathematical discoveries of the time, the foundations for science, statistics, and probability were laid, and humans could predict the future separately from the history and the myths of Gods. Numbers gave us the confidence that humans have power and that things are predictable. People started making algorithms to support decision-making based on the premise that the future could be calculated if enough data was available. This created a distance between people, their perspectives and values, and their decision-making. We also see this today, where numbers are perceived as objective and true. We tend to value more quantifiable benefits over non-quantifiable benefits. For example, physiological health and safety can be assessed using statistics and historical data. However, when we consider psychological health and safety, this might be less tangible, and it can be more difficult to argue for its value. Therefore, people who desire influence today try to quantify their points to impact decision-makers (Harari, 2019). Harari goes as far as calling this Dataism a form of religion.

In the 1930s, Chester Barnard took decision-making from public administration and introduced it to business, which, until then, had mainly focused on resource allocation and policymaking. Bernard and many more later theorists, thereby, laid the foundation for decision-making in management (Buchanan and O'Connell, 2006). In management literature and theories, decision-making centers mainly on describing the authority of those making the decisions, how they gain confidence as decision-makers, how they should act in a crisis, and what it takes to make an impactful decision. We often admire the brave: those who dare to make decisions, regardless of the outcome.

With the introduction of computers and, later, artificial intelligence (AI), we are beginning to rely more and more on data and algorithms for decision-making. AI can compute data extremely fast and process new knowledge and data, but we also need to consider the impact this technology can have on the system and the people involved. Running simulations using algorithms to compare alternatives gives people a feeling of all-inclusive objectivity, where all relevant data is incorporated, and the numbers are correct. This can cause people to feel a lack of agency, meaning they do not see how they can impact decisions. We lose insights into the input data and methods to reach the decision. Furthermore, it can be difficult to translate a certain decision into relevant actions if the decision process is not transparent (Arroyo et al., 2023).

Who are the decision-makers today?

Throughout history, we have expected some people to make decisions on behalf of the larger group. Think of kings, popes, and the village teacher. Today, we still consider decision-making a privilege that involves only a select few. We have management groups, boards, consultants, experts, people with authority, and other selected groups of decision-makers. In a group, they are considered to have the authority to make decisions and enable actions to happen. This expectation could be rooted in tradition, certain stereotypes, power, seniority, specialist knowledge, experience, etc. If in doubt, we look to the most powerful or respected person in the team or the meeting to make the decisions and to have the last word. Solo decision-making is not necessarily bad; it can be very efficient in times of crisis where it is crucial to act fast. Still, when considering decisions needing input from different perspectives and with vast consequences, individuals with authority must rely on more than just their own personal knowledge, experience, and preferences. It can, therefore, be worrying that we celebrate the courage to make bold decisions more than the ability to ensure a good outcome of the decision (Buchanan and O'Connell, 2006).

Consequences for generations to come

We are facing huge challenges with regard to climate change, biodiversity loss, and unequal development, and we are, as a global community, slowly starting to realize that past and present decisions are having significant impacts on future generations. We know we must change how we produce and consume, and we need to find a way to balance the climate, environment, economic, and social considerations. Despite facing serious consequences today, where extreme weather has become the norm, and we already see ecosystems collapse, we still do not know how to make changes in all layers of society. We see leakages from landfills contaminating groundwater; we see a worrying decrease in biodiversity due to accumulated micro-decisions favoring industrialized agriculture and urbanization. When we decide on a highway design, building a new hospital, or a school, the decisions will impact individuals and communities directly or indirectly for 50–120 years. Some wastes will stay in the landfill for hundreds of years, and some greenhouse gases will stay in the atmosphere for thousands of years. Who is the voice of the future? Of the planet? Who ensures we always consider their best interests when evaluating alternatives?

Often, we believe we know what is right and wrong or what the preference might be, and we jump right into a general conclusion and actions (Christensen, 2023). But is this the process leading to the best value for all? What is your strategy or the strategy of your company or team when it comes to decision-making? Do you know who has the authority? Or how to impact it? When people are asked: "How do you make decisions?," they stumble, think, and say that they are making it based on pros and cons, that one person is making the final call, and some can't explain how they make decisions. Why are we not questioning the decision-making process when we think about the consequences?

Do we have the right processes and methods to make collaborative decisions?

Not having a strategy, a process and a sound method makes it difficult to involve the right people in decision-making. Besides building on a history of solo decision-making, another reason for decision-makers to prefer to centralize the power could be the lack of knowledge of efficiently involving a group of people in deciding. With the strong focus we have on the outcome of decisions, we might have missed paying attention to process. Furthermore, if there is time pressure, many might assess that involving more people is too demanding and not worth the benefit of this extra engagement. Lastly, the lack of shared vocabulary and understanding of the process makes it difficult to involve and delegate, and, by nature, we steer away from conflicts in the group and avoid difficult discussions.

However, if we have a shared understanding of how to make decisions, we might be more inclined to include the right people and have a constructive discussion. If you have a plan for decision-making, you can question this without criticizing the people proposing certain alternatives. You can plan the process, consider who to include and what steps to take, and ensure these fit with other activities or schedules.

This book focuses on decisions requiring insights from more people or stakeholders, where conflicting values or perspectives need to be considered, and where transparency and acceptance are needed to allow the group to move forward with commitments and action. Having the right process, where those involved know when they can propose an alternative, they know how it is being assessed, and a decision is being made, will ultimately also lead to better-coordinated actions and strengthen the likelihood of seeing the expected consequences.

Key points

- We all make thousands of decisions a day, many of them unconsciously.
- Recognizing that we need to make a decision starts with being aware that we have alternatives.
- Decisions are made when we *act* on our choices, and the actions lead to consequences.
- Decision-making is a process, and we need to think about how we design that process.
- How we make decisions matters; this will affect the likelihood of arriving at the best decision.

- Historically only a few authorities made the decisions for society, people trusted in Gods or external forces.
- The people who make decisions today are changing. From solo decision-makers, organizations increasingly focus on being inclusive and diverse because it is better for all of us.
- Many of the decisions we make today have a big impact on future generations.

References

Arroyo, P., Shöttle, A., and Christensen, R. M. (2023). "A Shared Responsibility: Ethical and Social Dilemmas of Using AI in the AEC Industry." In V. Gonzáles, F. Hamzeh, and L.F. Alarcon (Eds.), *Lean Construction 4.0: Driving a Digital Revolution on Production Management in the AEY Industry*. New York: Routledge.

Baron, J. (2008). *Thinking and Deciding* (4th ed.). Cambridge: Cambridge University Press.

Bernstein, P.L. (1996). *Against the Gods: The Remarkable Story of Risk*. Hoboken, NJ: Wiley.

Buchanan, L., and O'Connell, A. (2006). "A Brief History of Decision Making." *Harvard Business Review*, January. hbr.org/2006/01/a-brief-history-of-decision-making

Christensen, R.M. (2023). "Different Strategies for Setting Requirements for Sustainability." Paper presented at IABSE Congress, New Delhi, India, September 20–22.

Christensen, R.M., Olsen, T., and Kean, A. (2022). "Lower Thames Crossing: Choosing By Advantages for a Sustainable Tunnel Solution." Paper presented at ITA-AITES World Tunnel Congress, WTC2022, Copenhagen, September 2–8.

Garvin, D.A., and Roberto, M.A. (2013). "What You Don't Know About Making Decisions." *Harvard Business Review*. HBR's 10 Must Read series. hbsp.harvard.edu/product/11367-PDF-ENG

Harari, Y.N. (2019). *Homo Deus: A Brief History of Tomorrow*. New York: Random House.

Howard, R.A. (1966). "Decision Analysis: Applied Decision Theory." In *Proceedings of the Fourth International Conference on Operational Research*. New York: Wiley-Interscience.

Kahneman, D. (2011). *Thinking, Fast and Slow*. London: Penguin Random House.

Kahneman, D., Lovallo, D., and Sibony, O. (2013). "Before You Make That Big Decision." *Harvard Business Review*. June 21. HBR's 10 Must Read series. hbr.org/2013/06/before-you-make-that-big-decision

Rhodes, P.J. (2015). "Decision-Making, Greek." In *Oxford Research Encyclopedia of Classics*. https://doi.org/10.1093/acrefore/9780199381135.013.2053

Suhr, J. (1999). *The Choosing by Advantages Decisionmaking System*. Westport, CT: Quorum.

2 Why decisions do not stick

Decision-making is about people. We make decisions individually, within a small group, or as a society, and our choices impact others on different levels without us always being aware of it. Decision-making can be very complex, and the more perspectives you include, the more challenging it will be to agree on the decision outcome. However, you will not get a broad understanding of the potential consequences without involving different perspectives. As explained in Chapter 1, a process is needed to support a constructive dialogue between people with different perspectives and help to create a shared understanding of the decision outcome. But before we show you how to set up the process, we will dive deeper into why decision-making often fails and what you need to be aware of to make decisions stick.

Not all decisions stick over time, as they might be changed or altered along the way. Changes in decisions are very common in many projects and organizations, regardless of the size. Sometimes, decisions require iterations as a deliberate strategy to make progress, learn, and improve the decision outcome. Sometimes, iterations occur incidentally and are uncontrolled, often resulting in a significant waste of time, energy, and money. Therefore, in this chapter, we will discuss the importance of communicating value and expectations transparently to make an effective decision, and discuss decision timing to avoid unnecessary iterations. We will also discuss the role of emotions, the importance of having a psychologically safe environment for decision-making, and the issue of biases and inclusion.

Communicating the why of a decision

A risk occurs if the purpose of the decision is not transparent and understandable to everyone. This especially happens when the people who decide and those who have to execute the decision are separated from each other. If the execution team does not understand why a decision was made, then there is a high probability that the implementation of the decision outcome will be faulty because no one takes ownership of it and everyone avoids action, or worse, people intentionally sabotage the decision outcome because they do not

DOI: 10.1201/9781003518310-3

agree with it. This often happens in large organizations with different hierarchical levels, where a decision has to be communicated through several levels. Annett's experience will illustrate the consequence of not communicating the why behind a decision.

This story explains one of the main lessons Annett learned when she was on a study trip in Japan in 2018. During a visit to a Toyota manufacturing plant, a very experienced top manager mentioned that the company had a high fluctuation of employees around 40 years ago in Japan. This was unexpected, as we normally associate Toyota and its approach to employee involvement with high transparency and satisfaction. He explained that the high fluctuation was a result of non-transparent decision-making within the organization. People on the operational level couldn't understand decisions made on a strategic level, based on the information given to them. The consequence was that employees left the company, to show their dissatisfaction with the situation. Searching for the root cause, the management had to learn that *it isn't enough to just communicate the outcome. They learned that they had to explain 'the why' of the decision.*

To be able to explain the why of a decision, you need to assess the decision from different viewpoints and develop clear arguments for the decision. If the outcome is not understandable, it can result in people questioning the decision and creating cognitive dissonance, where people might feel they have to act in a way that is in conflict with their values. Communicating the why behind a decision is crucial to enable efficient implementation and to ensure that potential improvement ideas can be included. This is especially a challenge for middle management in large organizations as they have to communicate the decision to the operations personnel so they can understand the intentions behind it and thereby implement it. If the operational personnel do not understand, they risk resistance and sabotaging behavior. Furthermore, the middle management has to communicate the why of a decision to upper management so they can trust it and provide formal acknowledgment.

Rushed decisions and their consequences

We often rush into a decision without having the full picture of the outcomes and consequences. We will illustrate this issue with this example.

Kate and Fernando were renovating their new house, and they started selecting the interior elements for the kitchen. One of the decisions was to choose the floor surface. There were several alternatives, such as tiles, wood, laminate, or vinyl. Their experienced kitchen interior designer, Susanne, suggested installing a wooden floor, a parquet. Kate and Fernando liked the visuals she made and agreed to this with great confidence—who doesn't like parquet? It looks nice, adds character to the room, and gives a feeling of an exclusive interior. So, perfect! They were happy that the decision was made quickly as more and more decisions were coming up for the other rooms.

Just as the contractor had finished installing the parquet floor, Kate realized that with their three children and their dog in the house, it might be more difficult to keep the surface looking nice. Fernando agreed that there would be wear and tear from toys and paws. Therefore, before installing the kitchen furniture, they decided to change the floor from parquet to tiles, as tiles are more resistant, and easier to maintain and clean. But changing to tiles means they had to take out the parquet, place a new order that would take longer to deliver, and find a new contractor to install the tiles. This was going to delay their house project by two months and increase the overall cost. The decision on the parquet did not stick. Why? One valid alternative was presented, and the decision was made in a rush to avoid short-term energy loss, but the consequences were long-term, as we see in this example.

We often make rushed decisions and later realize the unwanted consequences of our choices when our needs are not met. Rushing into a decision often has the opposite of the desired effect; the process takes longer due to missing pieces and potential iterations. Therefore, important decisions should get the time and focus required to think them through. Think about the tunnel project in Chapter 1. What would the outcome have been if the team had gone with the shorter tunnel without considering other alternatives? This could have created an unnecessary and significant negative impact on the environment and local community.

Another reason why decisions do not stick could be decision fatigue. This phenomenon occurs when many decisions must be made simultaneously, and the decision-maker is tired of deciding and just wants to finish. This also applies to our example. Think about Kate and Fernando having a decision session with Susanne where they have to decide on the kitchen counters, the appliances, and the lighting system, and at the end of the meeting, they have to decide on the flooring. This intense session led to a not well-thought-through decision about the flooring. We all have experienced something similar when we are negotiating a group decision over a long period. As time goes by, we get tired of negotiating back and forth. At one point, we want to be done and give in to move forward, but we may not like or may regret the consequences later on. Consequently, *decision fatigue results in neglecting to think all decisions through or consider all aspects.*

The right timing of a decision

Another important aspect impacting the decisions is the time at which a decision is made. When is the right time to decide? Making it too early can cause changes as new information might appear or already available information has not been sufficiently considered due to rushing the decision. Deciding too late usually results in frustration as people cannot proceed as planned, or opportunities might be missed. In this section, you will learn what it means to make decisions at the last responsible moment, when iterations of a decision are adding value or waste, and how important it is to communicate the status of a decision.

It is important to make decisions in a timely manner, as the following scenarios will illustrate. Based on our kitchen example, imagine Kate and Fernando changed their minds about the tiles before installing the parquet. They would still have to change the purchase of the product, but they would not have to go through the process of reworking the floor. This would not lead to a delay of two months as they could have ordered the tiles earlier and found the trade to install them to finish quicker. If they had changed their minds even before ordering the parquet, there might not have been any delay at all. *That moment when there are still no consequences related to your changes and all your options are still available is called the "last responsible moment."*

The above scenarios show us the importance of timely decisions in your projects. As Kate and Fernando made a decision after it had consequences to change, they had to accept rework. Therefore, the dimension of time is related to the consequence of a negative iteration. Negative iteration occurs when you must rework what you have already finished or repeat a task without having a better outcome, unlike positive iterations, which help you develop better ideas or mature alternatives for your decision (Ballard, 2000). Typically, you can have positive iterations early on in a project when you are still in discovery mode or early designing stage and because the impact of the change is low. Late changes usually lead to negative iterations.

It may be hard to distinguish when an iteration is adding value and when it is not. You are probably in a positive iteration when collaborating, learning, and rapidly changing your mind. Positive iterations add value to the project, and it is relatively easy to communicate the value of the change.

Working and deciding in isolation can lead to more negative iterations. If you decide in isolation and afterward communicate the outcome of the decision to the rest of the team, the team might provide you with new information that you need to consider. You might find that your decision doesn't work, but by then, everything is already set up, and the team is acting according to your original decision. You are now facing a negative iteration. This can be really painful because you and the team might have to redo work and face delays. This iteration is not adding value. You are not making the project better; instead, you have to communicate that it doesn't work and revise the decision. At this moment, the change will add negative stress on those who have to deal with the consequences as the rework often has to be done quickly. Often, negative iterations occur in non-routine tasks (Wynn and Eckert, 2017), as we could see in our kitchen example where the flooring decision was a once-in-a-lifetime decision for Kate and Fernando, and it was made too early without investigating the needs, resulting in a considerable rework.

The timing of the decision is, therefore, key. Making decisions too early also increases the likelihood of iterations with negative consequences (Terwiesch et al., 2002). If a decision is made earlier than needed, you risk that the input information status will change, meaning you learn something new. You will probably need to revise the decision based on the new information if you still have time to execute the decision outcome. Thus, if the project is still in the design phase, the consequences might be less serious compared to when things are already built, and the whole project suffers through the consequences of adjusting or starting certain aspects over from scratch. Therefore, decisions should be made as late as possible to generate enough information (Ward et al., 1995; Sobek et al., 1999). Toyota has a term for this concept, "Genchi Genbutsu," meaning make decisions slowly and implement fast (the Toyota Way, principle 13th).

However, you must also recognize that decision-making is always an attempt to predict the future, as we showed in the previous section, and you will never be able to cover every possible consequence in advance. You cannot wait forever and probably will never have all the information needed. So, ideally, you will make decisions at the last responsible moment.

One strategy to postpone a decision is by providing flexibility to adjust to multiple alternatives. For example, when hospitals are built, designers know it will take approximately 5–10 years from the conceptual design to the actual opening of the hospital, which means that you prepare the decision many years before you actually decide which equipment to buy. Hospital designers know that the owner wants to install the newest available equipment and that the technology will change rapidly. This can be a reason to delay a decision and wait until the last responsible moment. This, however, requires thorough planning as you will need to construct the building around this decision. For example, the project team can leave a hole in the side of the building to access the MRI room and plan to add the equipment with a crane shortly before the hospital opens. Thus, for some decisions, you must plan for flexibility, and the constraints around decisions must be communicated explicitly to protect the needed flexibility. Creating flexibility requires a well-defined decision-making process.

When discussing the right timing of a decision, the last responsible moment needs to be transparent to the decision-makers and the wider team. You can ask questions like: when do we need to make this decision, and what would happen if we made the decision later than that? When making a decision at the last responsible moment, it is also essential to continuously communicate the status of a decision to the team. The team needs to know when to expect a decision to be made, and when they can start acting on it. You risk experiencing conflicts if some team members believe the decision has already been made and start acting accordingly. At the same time, other team members might believe that an iteration is needed, and they will be waiting for new information. Here, there is no agreement regarding the status of the decision. Imagine you were in a meeting discussing a decision regarding the window frames for a new office building, and a person in the room decided to go with a specific wooden frame. He then started acting according to this and purchased the materials while another person was still exploring different options. This can happen when the parties do not clearly communicate and document the decision's status, especially when they lack a shared understanding of the process.

Understanding expectations and value

Who do you believe was responsible for the change in the decision of the kitchen floor? Susanne, the kitchen interior designer, because she suggested a parquet? Probably not. If Kate and Fernando did not understand or point out their needs, how could they expect Susanne to know what they needed? We often assume the other person knows what we want instead of explaining and discussing our expectations. The struggles with the kitchen floor could easily have been avoided by a conversation between Susanne, Kate, and Fernando to better understand their underlying needs. Susanne could have supported this by showing more alternatives and highlighting the differences between the alternatives. Kate and Fernando asked themselves privately: why did Susanne not tell us about the disadvantages of parquet? On the other hand, Susanne is only responsible for the kitchen interior and did not know about the three kids and the dog and, therefore, wasn't aware of the full needs of the family. When these conversations do not happen, people enter a bubble of assumptions, where both sides think that the other person knows what is not explicitly articulated. *Don't assume you know what people want; start by asking questions and articulate expectations.* On the flip side, be as clear as possible with your needs if you are the client.

There is often not enough discussion about the value upfront because the concept of value and values is often misunderstood. Here, we want to make an essential distinction between values and value. Values are the central beliefs and ideals that guide you in your life (Schwartz and Bilsky, 1987), for example, honesty, family, hard work, inclusion, integrity, kindness, enjoyment, impact, etc. Value, on the other hand, is more specific to the worth of what a thing or a person provides or brings to a situation (Drevland and Lohne, 2015), and it is judged from the client's perspective. For example, students value the space to play outside in a school, and teachers value a quiet space to prepare for their classes. Those two concepts are interconnected, as often your values are a baseline for interpreting how you value things in life. For example, if having a family is one of your values, you will consider that when choosing where to live and work. It is interesting to notice that when you have a group of people making decisions. Their values are probably different, and this will also translate into their interpretation of the value or worth of different alternatives. For example, a vacation in Las Vegas, Nevada, will be more appealing to someone who likes to party (connected to enjoyment as a value), vs. a vacation in Athens, Greece, could be more appealing to someone who enjoys visiting temples (connected to cultural appreciation as a value).

Diving deeper into the topic shows the complexity of decision-making within a group, especially when the participants come from different organizations and bring different values and interests to the decisions. You must realize the social motives (Collins and Guetzkow, 1964), and values within the group to make a decision that sticks. If the decision is made in accordance with the values, it is more likely to be implemented. However, suppose the decision is made in conflict with the values, and the reason is not communicated clearly (the why). In that case, it can, as described earlier, lead to cognitive dissonance, and the group is likely to actively or passively work against the agreed decision outcome.

Also, values change over time (Christoffersen, 2003), and if you do not realize these changes, it can result in decisions that constrain the future by not fulfilling or living up to expectations. For example, ten years ago, sustainability was rarely considered when designing a new office building. Now with the awareness of climate change, new values are emerging in our societies, and we need to consider these changes. Sustainability now has a different meaning than ten years ago, and we have to adjust our decision-making based on this change. This also means you cannot just copy and paste a decision you made a few years ago. The context will change, and so will values. We have to understand the purpose of a decision to align interests and to balance power relations and hidden agendas. Therefore, we must understand the values of the decision-making group and recognize that values change over time.

If we disagree on what value we are looking for without communicating about it, we may end up in negative iterations as we may change our minds, as described in the kitchen floor example. This example quickly shows that a decision does not stick if we do not understand the client's value. This kind of situation can lead to a tremendous delay in a project, but perhaps more damaging, it can lead to frustrations and conflicts in the team.

Considering emotions

We are not as rational as we like to think, and because we are human beings, our decisions are influenced by our emotions. It is in our nature to evaluate situations and interactions based on perceptions and emotions, whether we are aware of them or not (Barrett, 2019). Our emotions can provide relevant insights, which is a strength, not a weakness. When you work with people, you will have a different approach to the problems at hand depending on whether you or others feel anxiety, unfairness, enthusiasm, or joy. The issue is not your emotion, it is the unsaid statement behind it. By understanding and explaining why we feel a certain emotion, we can reveal an underlying insight that gives us a different perspective on a problem and will help us find a solution. Therefore, we as decision-makers need to understand that emotions will impact our decisions. We, as human beings, are not totally rational and impartial decision-makers. Our best strategy is to try our best to be aware of our emotions and the emotions of others.

So what could be the emotions Kate and Fernando felt when they rushed into the decision? They might have been overwhelmed and lost by the sheer number of decisions they have to make. Imagine they are living in a temporary apartment with a fixed termination date. This keeps them worried and anxious as it looks like they will never finish their house renovation on time. On top of that, they have too much on their plate and are stressed with their children and work, and simply do not make the time to think the decision through. In contrast, imagine if Kate's parents were living nearby in a big house, and they were happy to help out with the kids and they have the possibility to move in with them for as long as the renovation takes. This will probably make Kate and Fernando feel supported, safe, and excited about the new house. Think about how this emotional setting will impact the way they make decisions. In this scenario, they will have the opportunity to explore more alternatives and not rush into decisions.

It is important to understand how you feel when making a decision. Emotions impact how we process information, how we perceive risk, and provide insights into the underlying consequences we face when making decisions.

We do not pretend to be experts on emotions. There is much more to consider when it comes to emotions and decision-making, and we have only provided a brief overview here. Still, we wanted to make the point, given that in the construction industry many people try to hide their emotions or pretend we can be objective observers and decision-makers. Our goal is to ensure that you remain mindful of your emotions and the emotions of other people. Do not ignore emotions.

The impact of learning and a psychologically safe environment

Every decision leads to future consequences, and we cannot always predict how things will unfold. We judge based on what we know about the alternatives, our life experiences, our emotions, and the experiences of the people we are talking to. Our brain simply fills in the gaps in our knowledge based on expectations. Therefore, we do not necessarily know the truth of what will happen. Decision-making would be straightforward if we knew the exact outcome of choosing each alternative, but that is not the case. By involving more perspectives, you are more likely to recognize the blind spots in your assessments, and therefore, in this section, you will see the importance of aligning different interests by having an open conversation and discussing relevant facts.

If you discuss a decision with different people without changing your mind or learning anything new, you probably aren't listening well enough. There's always a bit of information you thought was certain, but through conversation, you find it wasn't cut in stone. This is a moment that can enable learning. Learning is: (1) to reconfirm your original expectation of a certain outcome, (2) to accept you have to change your expectation, or (3) to accept you have to act differently to expect a different outcome in comparable situations in the future (Christensen, 2008). So, in a situation of learning, you might find that you were wrong. There might also be information that strengthens your belief or perception, and you may have learned about a new argument. Most importantly, you also need to respect other opinions to be able to learn. Ideally, one person in the room talks about their perspective, and everyone listens. This continues until everyone has made their point. Then, a discussion can start, based on what was said. However, this is usually not the case. Think about your last group decisions. What happened? Who was the dominant voice in the room driving the discussions? Was everyone able to state their ideas? Was someone afraid to speak up? Did you get annoyed by someone else's argument, and, if so, why? You see, it is not easy to listen to everyone with the same attention. *Learning through conversation and sharing different perspectives form the first step to get to a decision that sticks.*

This requires an environment where people feel free to speak up, share ideas, ask questions, and express their concerns without being afraid of suffering negative consequences. This is known as psychological safety. As Harvard Professor Amy Edmondson points out, having a psychologically safe space where people can interact with one another is mandatory for authentic conversation (Edmondson, 2018). This doesn't happen on its own. People need training in collaborative decision-making to be able to make decisions in a safe and transparent way. We need to create a psychologically safe environment if we want to learn and have an authentic conversation.

In project teams, we can also see a difference in decisiveness in decision-making. There are people who like moving forward and want to make the decision alone, and there are people who are afraid or reluctant to make decisions. So, how can we get them into a good conversation to collaborate on making a decision? Besides creating a psychologically safe space, it is essential to have a shared understanding of the rules in the room, such as how much time you have set aside for different phases in decision-making. Here, a governance process will help to create the structure for the decision-making process (Schöttle et al., 2018). You can also level the needs of the different people by having a facilitator help a team through a decision. It also takes a lot of facilitation skills to keep people in the preparation mode to explore the decision to get the inside perspective and the data so as not to jump to conclusions too early. If you keep the group in the exploration phase longer, the actual decision-making can be done in a fairly short time.

It's very easy to challenge a decision based on opinions. But if you see and understand what went into the decision, the decision becomes more robust and is not perceived as an intuition, and people are more likely to stick with it.

The impact of biases

People make many decisions every day in both their private and work life. Some experts claim we make more than 30,000 decisions daily. For example, what to wear, what to eat, and who to talk to at the party. Of course, we cannot and should not spend too much time on every decision. That would make us not only very inefficient but we would also have problems getting out of bed because our cognitive ability would be overwhelmed. Luckily, our brains help us make fast and quick decisions based on our experiences and beliefs by structuring and organizing the information, but these shortcuts can result in biases. This is known as System 1 thinking, as described by cognitive scientist Daniel Kahneman (Kahneman, 2011). We have a built-in preference for this approach because it enables us to make quick decisions with little energy. However, there are also decisions where the consequences might impact others more than you and where you can't rely solely on your own insights, beliefs, and values. These decisions should be treated more carefully and considered, as described in System 2 thinking (Kahneman, 2011). In this section, we will see that biases consciously and unconsciously impact our decisions.

Biases are deviations from what is expected according to social norms (d'Orgeville et al., 2014). They are a tendency of preference and can result in an unfair judgment toward a person or an organization. We all have biases, but we are often unaware of them; they are unconscious (implicit bias) or conscious (explicit bias) and manifest in subtle or obvious ways. Therefore, a decision-making process needs to be defined in a way that helps overcome or minimize biases. We first need to understand how our brain processes information in group decision-making. Our brain does not analyze information precisely; our brain takes shortcuts based on our values, experiences, and preferences, resulting in subjectivity and sometimes systematic mental errors. This is called cognitive bias, and there are hundreds of cognitive biases, and researchers are still discovering new ones. There are different ways of categorizing cognitive biases. This book will explain the SEEDS Bias Model as it provides a good overview when making group decisions.

The SEEDS Bias Model classifies biases in similarity, expedience, experience, distance, and safety bias (Lieberman et al., 2015). *Similarity bias* occurs mostly unconsciously. It means that we prefer, for example, people who are like us over people who differ from us. This is why often people hire other people like themselves; unfortunately, by doing so, we exclude different perspectives that could help us decide based on a broader perspective. *Expedience bias* happens when we rush into a decision; we want to judge quickly, but in doing this, we fail to consider all the relevant facts. Consider our example where Kate and Fernando rushed into the floor decision and failed to consider how it would be used. Expedience bias has many subcategories, and we may not know all of them yet. Three very important subcategories are confirmation bias, anchoring, and availability bias. Confirmation happens when you select only the information or data supporting your beliefs. For example, reading only political news from your preferred newspaper and not checking other sources. This means you leave out information and reinforce your opinion on selected data. Referring to your decision based on the first received information is called anchoring. The result is that people jump to a conclusion without checking for additional information. Deciding based on the most accessible information is called availability bias. Continuing within the main categories, *experience bias* occurs when we do not consider that people see the world differently, and we assume that our perception is the ultimate and objective truth.

Distance bias occurs primarily in meetings where some people are in the room while others take part remotely as we tend to listen more to the input of the people near our physical space than people far away. *Safety bias* is the tendency of humans to avoid loss, and therefore, it often results in a sunk cost effect where we do not cut off the cost; we have invested in a project that doesn't work out and instead still try to turn around the situation by investing even more. One safety bias is the framing effect. Framing happens because information can be presented in various ways, and how it will be presented can significantly influence the information recipient. These are just some examples of cognitive biases, illustrating that our decisions are far from being objective. Therefore, it is essential to recognize and be aware of these biases.

So, what must we do to overcome our biases in decision-making? First, we need to create awareness that biases exist and no one is without bias. Second, we need to start the conversation about biases when we recognize them in discussions and meetings and by collecting data. Articulating biases requires a psychologically safe space in which to speak up and share experiences to create more awareness, develop strategies, and install tools to overcome or deal with biases. We will talk about this later in the book. It can also be helpful to think of the decision with distance, meaning that you are deciding for someone else or a different project. Finally, we need a structured way of making decisions that accounts for different problems to reduce biases.

Including multiple perspectives

We discussed the importance of defining value and expectations and communicating this to those involved in the decision-making process. We have discussed the impact of time, and the importance of sharing the decision status, and also that you should include different perspectives when deciding. Now we will discuss which people should be involved in the decision-making process. Including multiple perspectives is just one side of the coin, including the right people in the room is the other side.

Decisions often get questioned when different people are not included in the conversation. So, how do we know which people should be included? This could depend on the participants' availability, personality, competence, skill set, problem-solving ability, and conflict-resolution skills. We know that a diverse group will drive innovative solutions (Hunt et al., 2015; Bourke and Dillon, 2018), and therefore, diversity/equity and inclusiveness are key to driving decisions that need creativity.

The *Cambridge English Dictionary* defines inclusiveness as "the quality of including many different types of people and treating them all fairly and equally" (https://dictionary.cambridge.org/dictionary/english/inclusiveness). How often do you see decisions made in your company without including experts within the field? We need to understand if we have different perspectives in the room. Look around the room. Does everyone look like you, or do we have enough people with different perspectives on the problem in the room? A conscious effort must be made to check if the group is diverse regarding gender, knowledge, age, cultural and social background, etc., as well (Wasserman et al., 2008; Shore et al., 2011).

Including a knowledgeable person late in the decision often leads to a change in the perspective and questioning the decision. Furthermore, people with different opinions will challenge the decision and help you improve the outcome. However, acknowledging different opinions can enable a team to do more research and discover more facts before making a decision. For example, imagine Kate originally wanted a vinyl floor in the kitchen, while Fernando wanted to go with the parquet. By having a conversation regarding the advantages of each alternative, they might have found out what matters to them, and based on that, they could have asked Susanne which floor alternative best fulfills their needs. By doing so, they could have found that tiles would be the best alternative early on. Listening and acknowledging different opinions might lead to discovery of more alternatives and better alignment in values. This will help your decision to stick.

Decisions that have a long-term impact might need help from a facilitator. This is a person who is not involved in the project or decision but makes sure that the discussion in the group is level in terms of speaking up, hearing out, and aligning the different interests in the room. In this case, asking open questions is the best tool. The more different perspectives you include in the process, the greater the likelihood that the decisions will stick. For example, during a decision-making workshop for Deutsche Bahn for the Munich main station project, people with different perspectives were involved. They stated that the workshop was exhausting as discussions got intense, but they recognized afterwards that this detailed discussions directly helped the team to align on the decision outcome (Schöttle et al., 2023).

Think also about the tunnel project mentioned in Chapter 1, a decision method was implemented, and a team of facilitators was trained to facilitate the decision-making process. More different perspectives and inputs were valued when using a structured process. A defined method promotes inclusivity and a more open invitation process. By having a recognized structured process, the decision owners could also (better) resist the urgency to make the decision "on the spot" to meet a deadline (Schöttle et al., 2019). Leveling out these different factors is tricky but can be done with a structured and fair process that people trust.

Summary

We make decisions as humans, so a decision is never objective. We can agree on the decision outcome. Traditionally, in the construction industry, decisions are perceived as made hierarchically and based on a few influential voices. It is a part of the culture to make decisions quickly and presumably confidently. However, introducing a systematic and transparent process helps the team avoid biases and create psychological safety. Chapters 4 and 5 will provide a transparent system to help you overcome the main reasons why decisions do not stick.

Key points

- Decisions do not stick when people do not understand the why.
- Decisions do not stick when people rush them and do not understand the client's needs/expectations from different perspectives.
- Make decisions at the last responsible moment to avoid negative iteration.
- Create a space that is psychologically safe for people to speak up.
- Be aware that different biases impact decisions, and try to limit these biases by pointing them out in teams.
- Include people with different perspectives to increase the chance of decisions sticking.

References

Ballard, G. (2000). "Positive vs. Negative Iteration in Design." Paper presented at 8th Annual Conference of the International Group for Lean Construction. Brighton, UK, 17–19 July.

Barrett, L.F. (2019). *How Emotions Are Made*. Providence, RI: Book Festival.

Bourke J., and Dillon, B. (2018). "The Diversity and Inclusion Revolution: Eight Powerful Truths." *Deloitte Review*, 22(January), 82–95.

Christensen, R.M. (2008). "Development Practically Speaking: Learning Processes in the Danish Construction Industry." Department of Production, Aalborg University.

Christoffersen, A.K. (2003). "State-of-the-Art Rapport. Arbejdsgruppe Værdiledelse" [State of the Art Report: Working Group Value Management]. Byggeriets Evaluerings Centre, August.

Collins, B.E., and Guetzkow, H. (1964). *Social Psychology of Group Process for Decision-Making*. New York: John Wiley & Sons Inc.

D'Orgeville, C., Rigaut, F., Maddison, S., and Masciadri, E. (2014). "Gender Equity Issues in Astronomy: Facts, Fiction, and What the Adaptive Optics Community Can Do to Close the Gap." Paper presented at the SPIE Adaptive Optics conference, IV, 9148.

Drevland, F., and Lohne, J. (2015). "Nine Tenets on the Nature of Value." In *Proceedings of the 23rd Annual Conference of the International Group for Lean Construction*, Perth, Australia, p. 47.

Edmondson, A.C. (2018). *The Fearless Organization: Creating Psychological Safety in the Workplace for Learning, Innovation, and Growth*. Hoboken, NJ: John Wiley & Sons.

Hunt, V., Layton, D., and Prince, S. (2015). *Diversity Matters*. New York: McKinsey & Company.

Kahneman, D. (2011). *Thinking, Fast and Slow*. London: Penguin Random House.

Lieberman, M.D., Rock, D., Grant Halvorson, H., and Cox, C. (2015). "Breaking Bias Update: The SEED™ Model." 6 (November). Available at: davidrock.net/portfolio-items/breaking-bias…

Schöttle, A., Arroyo, P., and Christensen, R. (2018). "Demonstrating the Value of an Effective Collaborative Decision-Making Process in the Design Phase." In *Proceedings of the 26th Annual Conference of the International Group for Lean Construction*, pp. 899–909. doi.org/10.24928/2018/0500

Schöttle, A., Christensen, R., and Arroyo, P. (2019). "Does Choosing by Advantages Promote Inclusiveness in Group Decision-Making?" In *Proceedings of the 27th Annual Conference of the International Group for Lean Construction (IGLC)*, pp. 797–808. doi.org/10.24928/2019/0209

Schöttle, A., Guerreiro, R., Haubrich, A., Ditandy, M., and Behme, F. (2023). "Organizing the Argumentation for Changing the Delivery System Using Choosing by Advantages (CBA)." In *Proceedings of the 31st Annual Conference of the International Group for Lean Construction (IGLC31)*, pp. 990–1001. doi.org/10.24928/2023/0245

Schwartz, S.H., and Bilsky, W. (1987). "Toward a Universal Psychological Structure of Human Values." *Journal of Personality and Social Psychology*, 53(3), 550–562. https://doi.org/10.1037/0022-3514.53.3.550

Shore, L.M., Randel, A.E., Chung, B.G., Dean, M.A., Ehrhart, K.H., and Singh, G. (2011). "Inclusion and Diversity in Work Groups: A Review and Model for Future Research." *Journal of Management*, 37(4), 1262–1289.

Sobek, D.K., Ward, A., and Liker, J. (1999). "Toyota's Principles of Set-Based Concurrent Engineering." *Sloan Management Review*, 40(2), 67–83.

Terwiesch, C., Loch, C.H., and Meyer, A.D. (2002). "Exchanging Preliminary Information in Concurrent Engineering: Alternative Coordination Strategies." *Organization Science*, 13(4), 402–419.

Ward, A., Liker, J., Cristiano, J., and Sobek II, D.S. (1995). "The Second Toyota Paradox: How Delaying Decisions Can Make Better Cars Faster." *Sloan Management Review*, 36(3), 43–61.

Wasserman, I.C., Gallegos, P.V., and Ferdman, B.M. (2008). "Dancing with Resistance: Leadership Challenges in Fostering a Culture of Inclusion." In K.M. Thomas (Ed.), *Diversity Resistance in Organizations*. New York: Lawrence Erlbaum, pp. 175–200.

Wynn, D.C., and Eckert, C.M. (2017). "Perspectives on Iteration in Design and Development." *Research in Engineering Design*, 28, 153–184. https://doi.org/10.1007/s00163-016-0226-3

3 Focus on value and cost

In this chapter, we touch upon one of the more considerable roadblocks we see in making sound decisions: focusing too much on the cost and not enough on the value of the alternatives. Consider past decisions; how often do you make decisions focusing on cost? How frequently do companies, schools, or governments choose the cheapest alternatives? Did they understand the value of all the alternatives before making the decision? That is the type of question we will discuss in this chapter.

Far too frequently in the construction industry, we see people, project teams, and organizations making decisions based heavily on cost, but are they aware of the consequences of doing so? Often, people make decisions based on what they perceive to be the cheapest option, typically looking at the initial cost only. The predominant method for public owners' purchases worldwide is to mandate that the cheapest bid wins (Schöttle & Gehbauer, 2013; Schöttle et al., 2015). In this way, the public can see how much money the owner has spent, but they do not necessarily understand the rationale behind the decisions and the trade-offs the owner made. This procurement method applies to building public schools, hospitals, highways, ports, and airports worldwide, significantly impacting our quality of life, society, and the environment.

Both in the public and private sectors, there's a pressing need for transparency in decision rationale. Why are projects executed in a specific manner? This question often remains unanswered. Unfortunately, we don't always get the best possible outcomes for our communities using the resources at hand. We need to break this cycle. We need to encourage more questioning of the status quo in decision-making processes.

In this chapter, we challenge decision-makers to shift their focus from cost to value, and to consider the purpose behind their decisions. By clarifying the value stakeholders aim to achieve, decision-makers can lay a solid foundation for creating value. This shift in perspective can lead to better outcomes, inspiring decision-makers in the construction industry to make more informed and beneficial choices.

DOI: 10.1201/9781003518310-4

Separating cost and value

In Chapter 2, we described value as specific to the worth that a thing or a person provides or brings to a situation (Drevland and Lohne, 2015). In addition, in this book, we ask you not to think of value as an estimate of monetary worth, which is how the field of economics treats it. Here, we use value as the worth or usefulness of something irrespective of its cost.

When identifying or creating alternatives, ask what the value of each one of the alternatives is and focus on identifying trade-offs. The distinction between value and cost is essential, especially when making important decisions involving many stakeholders and when reversibility is difficult. If decision-makers do that, they will be able to choose the best value for the resources they have. We recommend that teams and organizations separate cost and value to make better decisions (see Figure 3.1).

Sometimes, we must make hard choices because the alternative that you or your team value more is also more expensive. Sometimes, we are tempted to go for the cheapest alternative, knowing it does not provide the best value (see Figure 3.2). In Figure 3.2 alternative A is less expensive and provides less value than alternative B. Have you ever bought something that you regretted getting? For example, have you ever purchased a cheap "alternative" phone charger that did not last or a cheap pair of shoes that hurt your feet? We have definitely made that mistake! Why did we not buy what we needed instead? Why not buy an original charger or comfortable shoes right away, even if they cost more? In the long run, what would then have been the better—and cheaper—alternative? We understand that we do not always have the money upfront to buy the most expensive alternative. For example, not everyone can afford a luxury car, and buying a cheaper car is understandable even when you know there are more expensive alternatives that you value more. However, we should consider more than just the initial cost in our decisions and keep those assessments separate

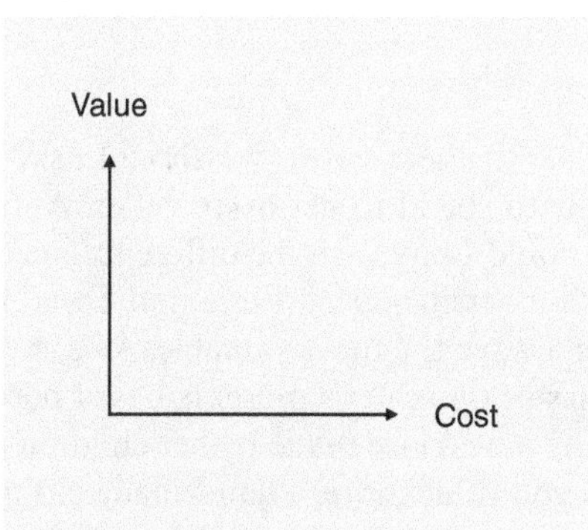

Figure 3.1 You want to see what value you are getting for the cost you are paying

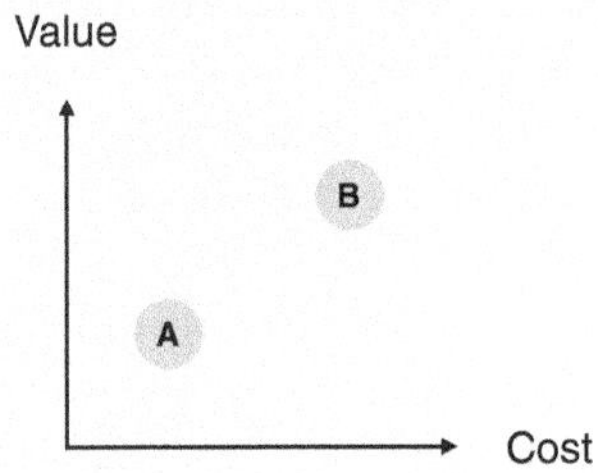

Figure 3.2 Example of a decision where the most valuable alternative is also the most expensive, such as in the case of a cheaper alternative phone charger that, over time, damages your phone battery vs. the original charger

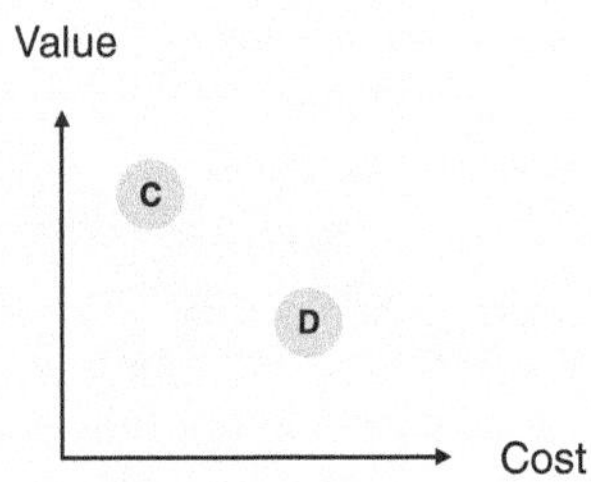

Figure 3.3 Example of a decision where the most valuable alternative is also the cheapest one, like a handmade necklace by your children vs. one they bought. It all depends on the context!

to make a sound decision (Suhr, 1999). In these cases, we should ask, how much more money am I willing or able to spend to get more value? Another question to consider here is, what else could I buy with the difference in cost?

We advocate that you decouple the assessments of value and costs. Let's not assume that the most expensive is always the most valuable. As a mother or father, you probably value your son's or daughter's presents based on their effort. A mother can be happier wearing a necklace made by her children than one they bought for her. Which would you value more, a hand-made gift made by your children or a bought gift that is more expensive? Again, value and cost are not necessarily correlated, but we often make decisions assuming they always are (see Figure 3.3). In Figure 3.3, alternative C is less expensive and provides more value than alternative D.

Questioning the relationship between cost and value is particularly relevant when we realize that some decisions have consequences beyond ourselves. Some decisions have an impact well past our lifetimes on the environment, a community, or a facility owner (Arroyo, 2014). However, we often make decisions as if cost is the only thing that matters. Let us examine some key points where we have a hyper-focus on cost and discuss the negative consequences of this approach. As stated, this happens too often in our lives, and we have been able to document this hyper-focus on cost in our research and our personal life experiences.

Hyper-focus on cost hides value

What are the consequences of hyper-focus on cost? In practice, when we decide with a hyper-focus on cost, we do not look at the whole picture; cost is just a part. Too often, other aspects of a decision, such as well-being, environmental issues, comfort, happiness, aesthetics, and impacts on future generations, are not or are only minimally considered. The cost does not represent all the consequences of our decisions. Be curious about the relationship between cost and value (see Figure 3.4)!

If we only focus on cost and obtaining cheap alternatives, we miss the opportunity to choose high-value alternatives. We need to understand how to evaluate value to choose what we value more. We need to remember the purpose of the decision, who is impacted, and how to create value for stakeholders. Think about all the effort that organizations and people put into quantifying the initial cost vs. the time spent understanding and defining value. People are

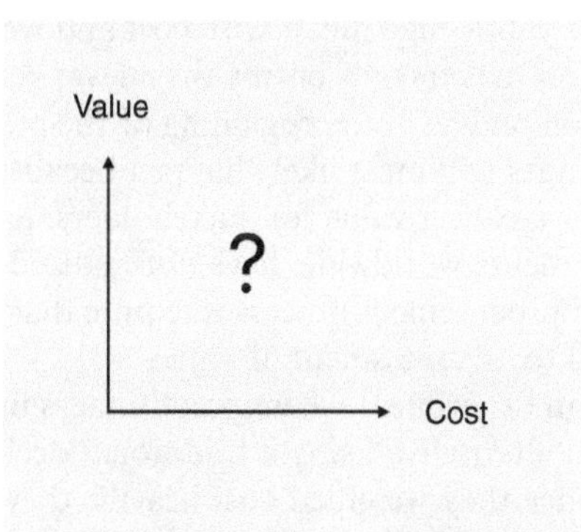

Figure 3.4 Be curious about the relationship between value and cost. Do not assume they are correlated, meaning more expensive is always better or worse; it all depends on the context of the decision

interested in knowing the cost of their alternatives and seeing if they can get it for less, but rarely do we spend the same time and mental effort defining and quantifying value. Often, we realize what we value too late in a decision after we have purchased or spent our money, time, and energy on an alternative that does not provide the value we wanted or needed. Chapter 7 explains how the decision-making method impacts whether or not decision-makers can differentiate cost and value.

The value of a decision depends on the alternatives considered and the context of the decision. To assess value holistically, we must consider multiple perspectives, including multiple, hopefully diverse, people in a decision. Including multiple perspectives that often conflict makes decision-making more complex and challenging, making many managers, politicians, and other decision-makers uncomfortable, understandably so. Many people do not know how to deal with conflicting opinions simultaneously, leading to analysis paralysis, meaning they cannot decide. Analysis paralysis happens when the decision-maker is overwhelmed by choices or has too much information. Fortunately, we can learn how to deal with multiple perspectives at once. Chapter 10 explains how to manage the decision as an organized, inclusive, and efficient process to avoid analysis paralysis.

Value for money

Do you want to use the least amount of money, or do you want to get the best value for your money? Focusing just on staying under budget can lead to poor decision-making. Hyper-focus on cost is an issue in many organizations; if we mindlessly choose the alternative that fits the budget, we lose sight of better alternatives within the budget or much better alternatives for slightly more cost. By focusing exclusively on the initial cost, we are not ensuring we get the most value for what we pay. A mistake we often need to correct is that when decision-makers want to include value, they mix it with cost and weight it very heavily. When doing this, decision-makers only ensure they never get the most expensive alternative. They stay on budget at the beginning of the project but not necessarily by the end, as changes will most likely happen because we have lost sight of the larger purpose—creating value for stakeholders. Sadly, many organizations, including governments worldwide, have not realized this misconception, and many bidding and procurement processes require that cost be weighted very high when compared to an assessment of value.

In Chapter 7, we present in detail an experiment where practitioners in the construction industry choose a design alternative using a traditional decision method that mixes value and cost. Since they weighted cost heavily, they got suboptimal results, meaning they did not choose alternatives that were better in value for the same cost as the alternative they had chosen (Correa et al., 2017). To illustrate this, in Figure 3.5, it would mean that they chose alternative C because it was within the budget line. However, alternative B was cheaper and better (but they did not choose it due to the decision method they were using).

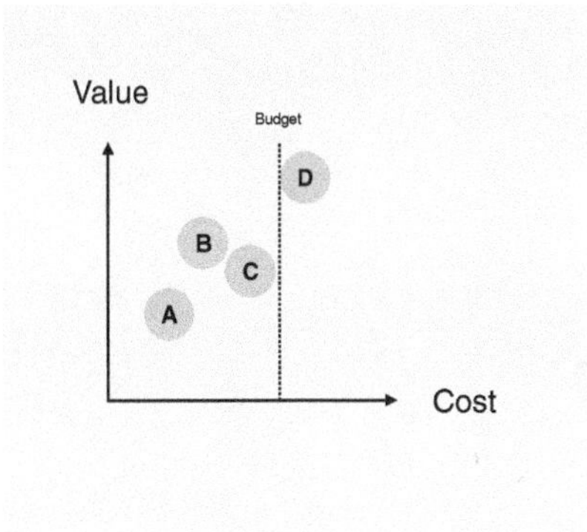

Figure 3.5 Example of a sub-optimization scenario. Choosing C because it is closer to the budget and missing B and D as better options, sub-optimizing value for the given resources you have

Alternative D was discarded even when it had significantly more value because it was slightly over budget. Practitioners who made that mistake did not see the graph of value vs. cost and could not visualize their sub-optimization.

This may seem ridiculous to you, but we have studied this sub-optimization phenomenon, and it is way more common than you would expect, especially when money cannot be moved across contractual or departmental boundaries in a project, organization, or government. Yes, cost is an important part of a decision, but we want to assess it separately from the value of the alternatives.

Assumptions in costs

We often assume that measuring cost is easier and more objective than measuring value. We would like to think that cost data is objective information; however, studies show that managers in several industries make important decisions about pricing, product mix, and process technology based on distorted cost information (Cooper and Kaplan, 1988). In the construction industry, projects use a budget to manage their resources, and typically, budgets are compartmentalized, hoping it will provide better control of the money spent. For example, a client can have $1 billion to build a new hospital, and that budget is further subdivided into design and construction; then, for construction, there is another subdivision for excavation, foundations, structure, façade, interiors, and equipment. How do we create budgets? Typically, budgets are based on estimates and assumptions about construction conditions, sequences, and production rates. However, we know we are often wrong about budgets (Flyvbjerg and Gardner, 2023).

Yet, we are willing to make decisions based almost solely on a cost assessment. Why do we keep trusting budgets (Figure 3.6)? Maybe because

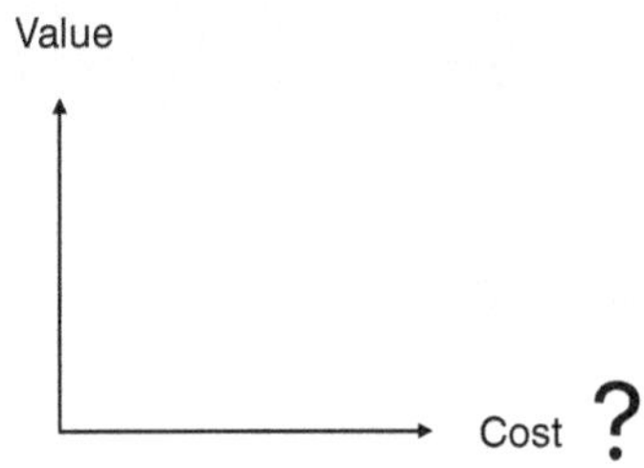

Figure 3.6 Do you trust cost?

the cost is relatively easy to understand. It is a number, and we have an agreed system to compare costs and access what we prefer. We can say the same about time—we know how a calendar works and can agree on a date across organizations. Hence, cost and time are perceived as objective—as facts—and relatively easy to compare when we have the numbers in front of us, and little can be argued based on these numbers. However, we know that budgets and schedules are often subjective estimations, and we do not always understand the assumptions used to produce them.

When it comes to value, we do not have the same accepted scale or even terminology to assess and compare different alternatives. We immediately perceive value as subjective and less critical. Different people prefer different things, and we, therefore, find it challenging to discuss and agree on them. Perhaps due to our need to belong, we shy away from discussions that can be difficult, particularly if we lack the language to discuss the issues. We need vocabulary, mental models, and processes to ensure a successful discussion around value. But this doesn't mean it is easier to determine the cost or duration of a service or product compared to determining the value. Research shows again and again how awful we are at predicting costs. The majority of projects overrun both schedules and budgets (Flyvbjerg and Gardner, 2023).

Cost is variable. Cost is not an intrinsic characteristic of an alternative; the cost of an item varies depending on where we buy it and how many we buy. For example, the cost of a light bulb will depend on the required quantity, the delivery terms, where it is purchased, and the market conditions. The cost of an alternative can be changed by negotiating with suppliers or changing the design.

Cost and time are also subjective assessments, as subjective as value is. Cost numbers include numerous assumptions and preconditions, risk assessments, and judgments. But as we have a uniform terminology to discuss them, we can better accumulate experiences and communicate our opinions. This could be the reason why we still prioritize discussion of cost and time over value, as we

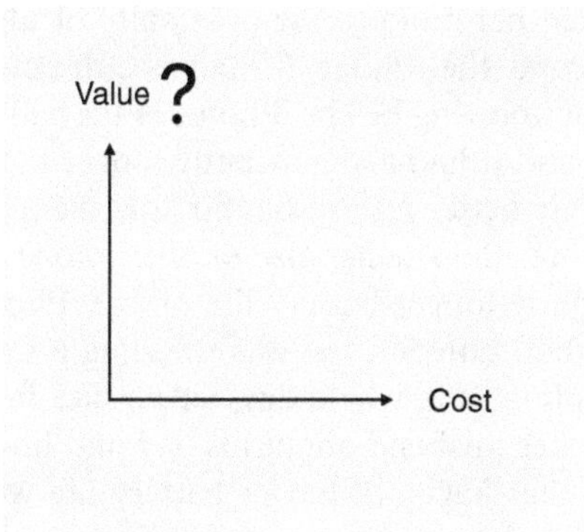

Figure 3.7 How do we understand the value?

find it easier to discuss—not because we are better at it or it is more important. When we make decisions based on costs before we understand the value of the alternatives, we will make poor decisions.

Evaluate value and then cost

We have discussed that we cannot make good decisions by hyper-focusing on cost, but instead, we recommend that you also focus on value first and then evaluate cost. We know that value is not easy to measure, and, often, it will take more upfront effort to discover (Giménez et al., 2019; see Figure 3.7). Value is context-dependent; you do not value things regardless of where you plan to use them and where you are in the world. For example, a liter of water when you are hiking in a remote mountain vs. a liter of water while you are working on your computer in your office does not carry the same value. There is no one-size-fits-all value system, and this can be tricky to understand for many people. In Paz's experience, we show a personal decision evaluating value first and then cost.

Phone example

A few years back, around 2016, Paz was a consultant and lost her smartphone. She was also a mother of two small kids, twins, and was traveling for work often. For many like her, this is something she had to replace right away as it was one of the primary forms of communication at work. So, she rushed to buy a new phone; at that time, she was a captured Apple consumer with a computer and all accessories from that brand; living in California, it was the norm. She knew she wanted a small iPhone (iPhone 7 at that time); it bothered her that the big phones did not fit in her pocket and felt too big in her small hands. So, she was happy to see that the smaller iPhone was also the cheapest alternative. She was about to buy it when she had to stop herself to consider her options carefully. When she opened her mind to the possibility of all the available alternatives (OK, just one more, the iPhone 7 Plus), she discovered that she was making trade-offs without knowing it. The iPhone 7 Plus had a 2x optical zoom (the iPhone 7 did not), and it had a significantly longer battery life (7 hours more); it also had a slightly better resolution. So now she had to ask herself what she really valued more: the smaller size of the iPhone 7 or the 2x optical zoom and the significantly longer battery life of the iPhone 7 Plus. When she looked at her life at that moment, she was traveling a lot for work and usually was checking the phone the whole day, letting her family know how she was doing and asking her husband about his day and how he was doing with the twins. She knew that having a longer battery life would allow her to have the peace of mind of being able to communicate without worrying about charging her phone during the day, and that was the most valuable advantage. Then, she thought about what she valued most next, and she realized that having a 2x optical zoom vs. none was great; she would be able to take better pictures of her kids growing up at her fingertips. Finally, she did value the smaller size of the phone, but it was way less important to her than the 7-hour longer battery life and having the 2x optical zoom. Now, the only question she had left was, was she willing to pay more for the iPhone 7 Plus? Her answer was yes, because she knew what she valued, and it was easy to choose after she had clearly seen the trade-offs.

We are very fortunate that we can apply this way of thinking in our lives, and we can teach and do research about it. Paz applied a system called Choosing By Advantages (CBA). We will explain how to apply it consistently in Chapters 5 and 6. For now, notice that she started by differentiating her alternatives and identifying the trade-offs she had at hand to make the decision. When we see how context-specific decisions are and identify value-based trade-offs in this manner, decisions become easier to make. Also, notice that she focused on the cost after she understood the value of the alternatives.

In this phone example, you may think it is easy to decide on your own as you only have to reason the decision to yourself, but often, we are not alone in a decision-making process. Whether you decide where to spend the next family vacation or if you have a decision in your company or project, the most important decisions are group decisions, and these decisions are more complex as different perspectives need to be considered. As you can imagine, as soon as you add more people to the decision, defining value requires making assumptions public and testing them. In Chapter 6, we present an example of how to do this in an example where many stakeholders are engaged in defining value to choose a mechanical system in a building.

Assess value against both initial and life-cycle cost

Understand the life-cycle cost as well as the initial cost. Most decisions have both immediate and long-term consequences. For example, in the construction industry, what we build will be there for many years; typically, designers plan for at least 50 years, and the current trend is to extend buildings' life-cycles as much as possible, including renovations.

In many cases, decision-makers can dismiss more innovative or sustainable ideas too early because they are hyper-focused on the initial cost. A manager can abandon further exploration of ideas that do not fit the set budget, missing long-term opportunities to create value. For example, installing a photovoltaic electric system on the roof of a building will have a higher initial cost than a more traditional natural gas-based energy system. However, we do not have to pay to capture solar energy; we must keep paying the utility company for gas in a natural gas system. Therefore, a photovoltaic electric system will be cheaper in the long term than a natural gas energy system.

Analyzing both initial and life-cycle costs is essential when considering sustainability in our decisions. We want to look at the value of the alternatives compared to both the initial cost (Figure 3.8) and the value compared against the life-cycle cost (Figure 3.9). In Figures 3.8 and 3.9, Alternative A has less value than Alternative B and is cheaper in the initial cost, but more expensive when we look at life-cycle cost. Alternative B is more valuable and more expensive in the initial cost but cheaper in the life-cycle cost. Then, decision-makers can evaluate if it is worth investing in a higher initial cost to choose Alternative B, knowing that in the long run, it will be better and cost less.

We hope by now you can see that to make better decisions, we want to separate value from cost and allow for a different type of reasoning. Decision-makers should ask, what value can I get from the initial cost and how does that compare to the budget or resources that I have available now? Additionally, when decision-makers evaluate value against the life-cycle cost and discover an alternative that is better in the long term (has more value and saves money in the long term) but has an additional initial cost, decision-makers can ask how to get the additional resources in the short term to pay for the extra initial cost to obtain the value they want. On the contrary, if we keep the status quo and

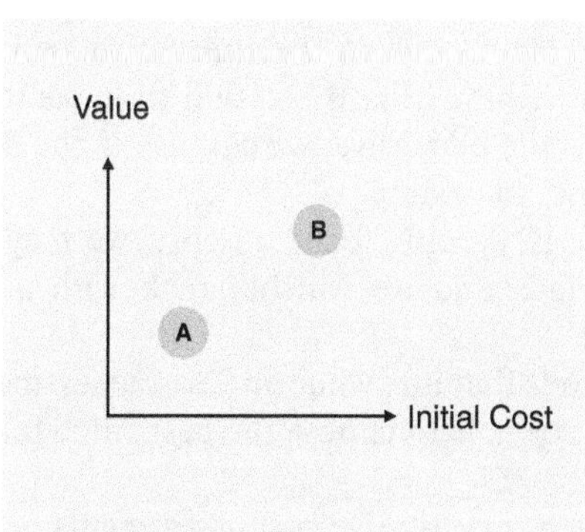

Figure 3.8 Value vs. initial cost

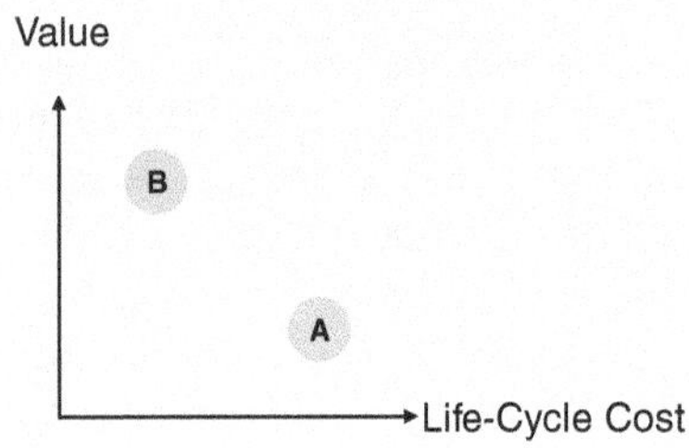

Figure 3.9 Value vs. life-cycle cost

keep making decisions by hyper-focusing on initial cost rather than on value, we're preventing individuals, teams, organizations, and governments from seeking the funding or budget they need to choose more valuable alternatives.

Summary

The main point is that we cannot just hyper-focus on cost; sound decisions that stick over time call for incorporating multiple perspectives, such as quality, productivity, safety, and environmental and social impacts, and for that, we need to consider multiple criteria in the decisions. Decision-makers need to be able to assess value holistically and inclusively, yet be able to manage a process that delivers decisions on time and helps evaluate initial and life-cycle costs.

Key points

- Cost is different from value, and we should assess them separately. Sometimes, we can make an easy decision when the alternative we value more also costs less. Other times, we have a hard decision to make if the alternative we value more is also more expensive; we can ask if the difference in value is worth the difference in cost.
- Hyper-focus on costs hides value; if we only focus on cost, we may not understand value until it is too late, and we will be stuck with a bad decision.
- Do not use decision-making methods that mix value and cost assessments; keep them separate, or we will not be able to choose the best value for the money we have.
- We often confuse transparency with quantitative cost assessments, which are subjective. Both cost and value assessments likely involve assumptions; we want to share and test those assumptions.

- We can explore what we value by seeing the trade-offs based on a specific decision context (see phone example). Focus on evaluating the value and then consider the cost differences.
- Differentiating initial from life-cycle cost is relevant to making long-term decisions; this is closely related to innovation and sustainability.

References

Arroyo, P. (2014). "Exploring Decision-Making Methods for Sustainable Design in Commercial Buildings." Doctoral dissertation, University of California, Berkeley.

Cooper, R., and Kaplan, R.S. (1988). "Measure Costs Right: Make the Right Decisions." *Harvard Business Review*, 66(5), 96–103.

Correa, M.G., Arroyo, P., Mourgues, C., and Flager, F. (2017). "Comparing Choosing by Advantages and Weighting, Rating and Calculating Results in Large Design Spaces." In *Proceedings of the 25th Annual Conference of the International Group for Lean Construction*, pp. 259–266.

Drevland, F., and Lohne, J. (2015). "Nine Tenets on the Nature of Value." In *Proceedings of the 23rd Annual Conference of the International Group for Lean Construction (IGLC)*, Perth, Australia, p. 47.

Flyvbjerg, B., and Gardner, D. (2023). *How Big Things Get Done: The Surprising Factors That Determine the Fate of Every Project, from Home Renovations to Space Exploration and Everything In Between*. Signal.

Giménez, Z., Mourgues, C., Alarcón, L.F., and Mesa, H. (2019). "Proposal of a Model for Measuring Value in the Design Process." In *Proceedings of the 27th Annual Conference of the International Group for Lean Construction (IGLC)*, pp. 49–62. doi.org/10.24928/2019/0196

Schöttle, A., Arroyo, P., and Bade, M. (2015). "Comparing Three Methods in the Tendering Procedure to Select the Project Team." In *Proceedings of the 23rd Annual Conference of the International Group for Lean Construction (IGLC)*, pp. 267–276.

Schöttle, A., and Gehbauer, F. (2013). "Incentive Structure in Public Design-Bid-Build Tendering and its Effects on Projects." In *Proceedings of the 21st Annual Conference of the International Group for Lean Construction (IGLC)*, Fortaleza, Brazil, July 31–August 2.

Suhr, J. (1999). *The Choosing By Advantages Decisionmaking System*. Westport, CT: Quorum.

4 Changing bad habits

In the previous chapters, we have discussed the need for a decision-making process and mentioned that decision-making methods matter. Different methods result in different decision outcomes and, therefore, different consequences. Other than the problem presented in Chapter 3, where we tend to make decisions based primarily on initial cost, we will discuss other pitfalls of traditional decision-making methods. We call these "bad habits," i.e., regular tendencies or practices that are hard to give up as we make decisions. We need to stop these bad habits if we want to make better decisions. When we try to assess value, we typically fail to go into the level of detail that allows us to learn about our preferences or the preferences of those affected by our decisions.

This chapter presents essential ideas from the Choosing By Advantages (CBA) system, developed by Jim Suhr (1999). Working as a US Forest Service engineer, he noticed that getting alignment on decisions is difficult and that traditional methods, such as listing pros and cons, Weighting, Rating, and Calculating (WRC), and other decision methods, were not sound. Traditional methods didn't help to analyze the differences among the alternatives and understand value. He dived deeper into how decisions were made, and along with several supporters, using the theory of general semantics, he developed the CBA system to make sound decisions that facilitate the involvement of several stakeholders (Parrish and Tommelein, 2009). Along with developing the CBA system, Suhr introduced definitions, mental models, principles, and a suite of methods to support decisions of different complexity. We will dive into the actual methods he developed later in the book. For now, we focus on some of the principles and definitions. In addition to Suhr's ideas, we also include other perspectives in this book by looking at decision-making as a process. We discuss how to manage the decision-making process more efficiently and consider the human aspects, such as acknowledging that we are not solely rational beings. As we have mentioned before, we have biases and emotions that influence our decisions (Kahneman, 2011), and we often take the easy way (System 1 thinking), which is based on our previous knowledge instead of having a conversation to learn and innovate (Spetzler et al., 2016).

DOI: 10.1201/9781003518310-5

Bad habit 1: Rushing people to make decisions without training

In many organizations, leaders are expected to make decisions or push others to make decisions without paying attention to the process and method used for decision-making, assuming that people already know how to do it. Jim Suhr (1999) states the following principle: *Decision-makers must learn and skillfully use sound methods*. This principle reminds you that you will need to learn sound decision-making methods. This is learning a new skill, such as learning to ride a bike for the first time; you need to invest time and practice. Do not expect others to just jump to CBA and be able to follow you; everyone needs time and practice. It is normal to go back to the unsound methods you already know, so keep this in the front of your mind.

CBA is not a mechanical or normative method where you just follow the principles and automatically arrive at the right decision. You need to understand the underlying principles and intentions to adapt to the specific context of the decision. In 2019, we researched what it takes to teach this method, and we noticed that even when we showed people how to decide with CBA, they would not be able to make a decision on their own without further practice and coaching (Arroyo et al., 2019).

When we teach CBA, we use Gloria Flores's (2016) ideas from her book, *Learning to Learn and the Navigation of Moods*. Gloria explains which moods are needed to develop a new skill using Dreyfus and Dreyfus's (1980) skill acquisition model that describes the evolution of behaviors starting from a beginner, then novice, competent, proficient person, until achieving an expert and finally mastery level. For example, to be a beginner, you must follow the rules, and, according to Gloria Flores (2016), the typical moods that allow you to do that are wonder, trust, and confidence. Wonder about the topic; there is something here for you. Trust in your ability to learn and confidence in your instructor. On the other hand, moods that will prevent you from learning will be confusion, frustration, impatience, and insecurity. If you are experiencing any of those moods, you will probably just drop the book here. In face-to-face training, we encourage people to ask questions if they are confused, take a break if they are frustrated, and come back later. Here, we can just encourage you to keep an open mind and bring your mood of wonder; imagine all the good decisions you and your team will be able to make! Trust you are able to read the next example, and this will make more sense. Finally, confidence, we know this method works. If you made it to this point in the book, we deduce you already believe us on that. So instead of just rushing a team to make decisions, encourage them to learn a sound method, such as CBA, to make decisions together. By the end of this chapter, you will understand why we state that CBA is a better decision-making method.

Bad habit 2: Generalizing when trying to understand the value

One of the biggest mistakes we make when making decisions is to try to weight or understand what is more important in the abstract (or with a high level of abstraction). How many times are you asked in your project, company, or personal life to make a generalization in order to inform a decision? For example, do you prefer fruits or vegetables? Do you care more about safety than productivity? Do you think women are more competitive than men? Do you think Europeans are more serious than Americans? None of these questions are sound! If you can answer them without research or jump to a guttural answer, you are just used to basing your decisions on prejudice. We can fall into making generalizations, as we assume our preferences can be applied to a wide spectrum of situations and alternatives. But instead of making decisions easier, generalizations make decisions less transparent and harder to communicate. To make sound decisions, you have to unlearn this habit and start asking a lot more questions.

What questions should you ask? First of all, you need to understand the purpose and the context of the decision. For example, if you have to choose between a fruit and a vegetable, you have to say "It depends," and ask:

1. *What is the purpose of the decision?* For example, you want to know if this is a dessert after-dinner or lunch option. Is it only for me, or am I choosing for a large group of adults? Are you eating on the go or in the office?
2. *What are my alternatives?* What type of fruit and vegetable are we talking about? Am I comparing an orange to a carrot? Or a banana versus a broccoli brunch? Avoid generalizations that use a high level of abstraction, such as comparing fruits vs. vegetables. We want to be more specific when identifying alternatives, using a lower level of abstraction, such as comparing a medium-ripe banana vs. a cup of fresh broccoli (Figure 4.1, this example is inspired by Jim Suhr's book).
3. *What factors are relevant to compare the alternatives?* Now that you understand the context and the alternatives, what factors will help me

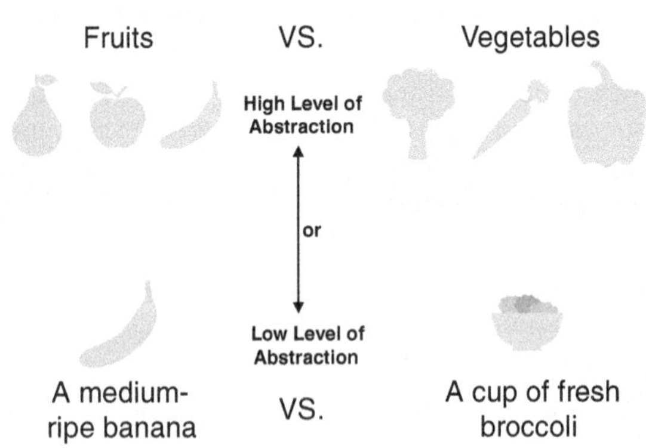

Figure 4.1 Example of abstraction (fruits vs. vegetables)

differentiate between them? Is it calories, taste, quantity, fiber content, or how fresh they are? Maybe all of them?

4. *What criteria can I use to judge each factor?* For example, using calories as a factor, you now have to make explicit what is better or a must-have. Is it more calories better or less calories better? Is there a minimum or maximum amount of calories you need in this decision context? The answer may be different if you are hiking or on a diet. This is not always an obvious question, and you will also find that different people have different criteria. You have to ask this question for every relevant factor that differentiates the alternatives to make a decision.
5. *What attributes do I know about the alternatives so I can judge them in each criterion?* I can only ask this question if I know the alternatives, factors, and criteria relevant to making a judgment. Continuing with the calories factor, assuming we are comparing a medium-ripe banana versus a cup of fresh broccoli, the questions here would be: How many calories in a medium banana? And how many calories in a cup of broccoli?
6. *What are the advantages of the alternatives?* Here, you see how the alternatives differentiate. If we continue with the calories, what is the beneficial difference in terms of calories? Let's assume you say more is better. You have to apply the criteria to know what the advantage is. The banana has 120 calories, and the cup of broccoli has 30 calories. So, the banana has the advantage of 90 calories more (120 minus 30).
7. *How do I value those advantages?* Once you are able to understand the beneficial differences or advantages of each alternative, you can judge which one of those is more important for the decision context. We will show you how to assess the importance of the advantages in Chapter 5.

According to Jim Suhr (1999), *Decisions must be based on the importance of advantages*. This means that we first have to understand the advantages of the alternatives we are comparing and then assess how valuable those advantages are. He called this the fundamental rule for sound decision-making. We agree that this is a big differentiator for CBA and probably the most important principle of sound decision-making. To make decisions based on the importance of the advantages, you need to state what the advantages are, and to do that, you need to know what are the alternatives that you are comparing, what factors you will consider, what criteria you will use to judge them, and how they perform on each criterion, which is called attributes. Therefore, you need to collect reliable information that is as close to factual as possible.

The easiest way to understand how this differs from traditional decision-making methods, such as the weighted sum (also known as WRC) or the Analytical Hierarchy Process (Saaty, 1988), is by providing a quick example. Imagine you must choose a car; safety and aesthetics are two factors to consider. Traditional methods, such as the two previously mentioned, will ask you to weight factors, which is a higher level of abstraction than weighting an advantage. You need to answer what you care more about, safety or aesthetics. However, that is an unsound and irrelevant question. What is relevant is how the cars you are considering differ regarding safety and aesthetics. Where do you get the most value? What are the most important differences? Suppose all the cars you are considering buying are equally safe but vary widely when it comes to aesthetics. In that case, the advantage in aesthetics will be more important for this decision because there is no advantage of any of the alternatives regarding safety. In that case, the decision is based on the importance of the advantage and not on a generalization of what factors you prefer. Therefore, you must assess your decision at the right abstraction level, where you see the differences in a specific context.

Bad habit 3: Double-counting the same differences

Many decision-making methods advise you to create a list with *pros and cons*, *advantages and disadvantages*, or *strengths and weaknesses* of the alternatives you are comparing. However, if you do this, you are double-counting and risk getting a false impression of where you get the most value.

An advantage is a beneficial difference, and it is in favor of one alternative, and describing the disadvantage is not only unnecessary; it is also double-counting the same difference. For example, imagine you have to select a player for your basketball team. You may assess the relevant differences between those players, which leads you to some factors to consider, like height, speed, precision, capacity to play in a team, etc. Now, if you take one of the factors, such as height, you can measure the differences between the two players. In this case, we'll say you must choose between Michael Jordan and Yao Ming (see Figure 4.2, inspired by Jim Suhr's book). Michael Jordan's attribute when it comes to height is 1.95 m; on the other hand, Yao Ming is 2.28 m. In this context, when you consider the criteria, taller is likely better; therefore, the difference, which is 33 cm, is an advantage for Yao Ming. You could also think

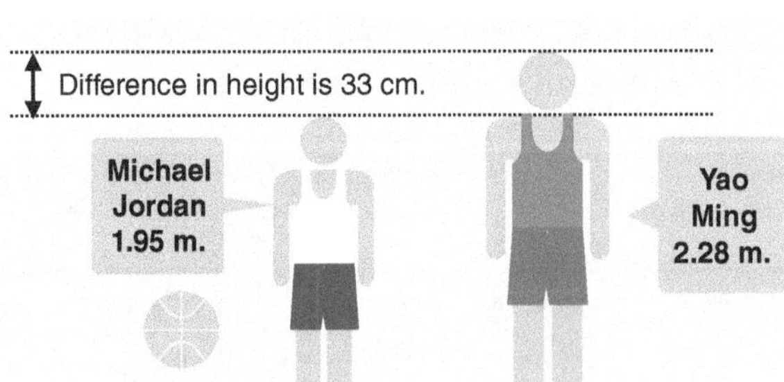

Figure 4.2 Height: Michael Jordan vs. Yao Ming

Table 4.1 CBA vocabulary

Term	*Definitions*
Alternative	Two or more people, things, or plans from which one is to be chosen
Factor	Element, part, or component of a decision
Criterion	A standard on which a judgment is based. It can be a *must-have* criterion (i.e., an alternative must comply with maximum or minimum criteria; otherwise, it is discarded) or *want-to-have* criteria (e.g., more or less is better)
Attribute	Characteristic, quantity, or quality of one alternative
Advantage	The beneficial difference between the attributes of two alternatives (one of which is the least preferred)

Source: Adapted from Suhr (1999).

about the disadvantages, and in this case, Michael Jordan will have a disadvantage in height, and the disadvantage will also be 33 cm.

As you can see, if you assess both as an advantage for Ming and a disadvantage for Jordan, you are now double-counting on the same difference. Therefore, you either choose by advantages or choose by disadvantages. Here, we encourage you to use the first so you only look at the beneficial differences between the alternatives. If you think about a disadvantage first, you can always reframe it as an advantage of another alternative, so don't worry if those come first to your mind; you can still use that information.

We hope this clarifies that pros and cons, advantages and disadvantages, and strengths and weaknesses are just ways of double-counting the same differences. Please do not use these methods now that you understand you are double-counting. What we're trying to do in CBA is to make decisions based on a context-based understanding of the differences between the alternatives. Once we understand the advantages of each alternative and their magnitude, we can understand how relevant they are for the decision that you're making in the context of the decision.

Remember these CBA definitions shown in Table 4.1.

Bad habit 4: Assuming preferences are linear

When using traditional decision-making methods, you will rate the performance of one alternative. Let's say you want to rate the performance of speed and height for choosing your basketball player. We would assume that faster is better and taller is better. However, is every increment in speed and height worth the same? There is no natural law telling this as a rule, so we typically create a linear scale to assign proportional weights to all sorts of things (Figure 4.3, bottom left; in this case, value is a linear relationship with

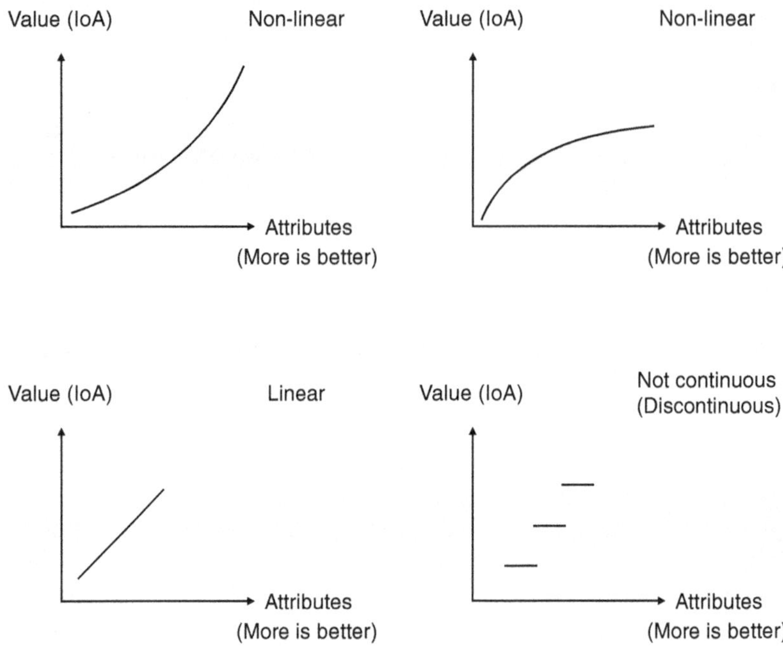

Figure 4.3 Example of preference curves. Note that not all preferences are linear
Note: IoA= Importance of Advantages.

alternatives' attributes). When we do that, we say that all increments within one factor are worth the same, thus assuming a linear preference curve. However, we know that is not true; we need to look at the whole picture or the bigger system to make a judgment call. There must be a point when being taller does not provide more value for your basketball team (Figure 4.3, top right; in this case, value is a non-linear relationship with alternatives' attributes on height). Now think about speed; when a player is making a small increment at the very bottom of the scale, going from slow to slightly better, the worth of that change is not as valuable as an increment in speed when the player is already really fast (Figure 4.3, top left; in this case, value is a non-linear relationship with alternatives' attributes on speed).

When we think about building interior walls in a hospital building, we may think that *faster installation is better*. However, that does not mean that every day saved is worth the same. We do not want to go faster than the rest of the trades. If the next task cannot be sped up, you will create waiting time and save no time regarding the overall schedule. Imagine the equipment will not arrive faster, and then you do not want to go beyond a certain date. On the contrary, if the project is late, the days that can be gained to get the project back on the schedule may be worth more than the days past that point (Figure 4.3, bottom right; in this case, value is a discontinuous relationship with the time attribute). Value preferences can have a minimum and a maximum, too.

In summary, when we think about how we value or prefer the attributes of an alternative, such as speed measured in days in building walls or meters/second of a player running, it is likely not linear, and a curve may be a more accurate way of reflecting our preferences. So, when you compare advantages, you are actually looking at the whole picture and making an assessment of what advantages are more worthwhile and not assuming linear trade-offs.

Bad habit 5: Confusing opinions with facts

We know there is a difference between opinions and facts, but they can easily be mixed up in real-time conversations. In his (2012) book, *Conversations for Action and Collected Essays*, Fernando Flores writes about using "action language" to maintain a culture of commitment in working relationships. Expanding on the linguistic action work of Austin (1975) and Searle (1969), Flores defines six basic speech acts: declaration, request, promise, offer, assessment, and assertion. Understanding the nuances of each speech act can explain dysfunction in communication and shed light on what actions to take. In the particular case of decision-making, we see a big confusion between two of those speech acts, assessments (opinions or stories) and assertions (factual statements).

When a speaker shares an assessment, it can open a new possibility or prepare for action in the future (Flores, 2012). This assessment can be grounded or ungrounded. For example, the basketball team coach said, "Speed is the most important trait of a basketball player." This is an opinion, not a fact (an assessment, not an assertion). It can be grounded on the experiences of the coach, but it would be really hard to prove it right or wrong. An assessment like this, confused as a fact, can influence the decisions of other people; a player may focus on only training for speed, and another player may say, "I am not good at speed; I may as well give up my basketball career." When a speaker is stating an assertion, the speaker reports facts and is prepared to offer evidence. For example, "Most basketball players in my team are taller than 1.85 m." This statement can be proven right or wrong.

When you make decisions, you need both speech acts, assessments and assertions. But when you make assessments, you want to test them against the group's dissenting views and share your standards or criteria when you are presenting your opinions. For example, you can have an assessment that a player is tall, but what is your standard to say that? Who are you comparing against? Is that standard shared with the rest of the team? You want to make your assessment as explicit as possible. To do that, you can use relevant facts;

in this case, state what are the heights of the players you are comparing against to say a player is tall.

Children's hospital example

In practice, it is really hard to come up with facts, and people often confuse their opinions with facts, especially in hierarchical organizations, as many construction projects traditionally are. For example, one time, Paz was helping a team make a decision because they were stuck. They needed to decide which mechanical system to buy for a Children's Hospital project. The mechanical engineer was clear on what the attributes of each of the systems were; on the other hand, the general contractor was clear on what the cost of each of these alternatives was. However, the owner could not make up his mind because he could not understand all the implications of the decision. So, Paz led a workshop to understand the advantages of each one of the alternatives. The team did the research with the mechanical engineer and stated clearly what the advantages were in terms of air quality performance, maintenance requirements, and ease of building, among other factors. Then, the stakeholders joined for a meeting; including the owner, the mechanical engineer, the general contractor, the facility managers, and a group of doctors and nurses so they could tell firsthand how they would evaluate those advantages.

As the team was in the process of deciding which advantages were more important, one of the nurses made a very passionate speech about the importance of air quality in a hospital. It went something like this: "The most important thing in this hospital is air quality. We are treating sick children here, and we cannot afford to have bad air quality that will lead to respiratory diseases spreading across the building and creating a catastrophe in case of this happening." So, what is this statement? Is this an assertion (a fact), or is this an assessment (an opinion)? Should we choose the alternative with the best air quality and forget everything else? Well, Paz realized that it was an assessment; even when it is grounded in a valid concern, it is not a factual statement based on the relevant facts for the decision we were making.

Because we were using CBA, we can ask how different the alternatives are performing when it comes to air quality. And because the team did the research first with the help of the mechanical engineer, everyone could see that all the alternatives considered by the team were great in terms of maintaining a high air exchange rate. Therefore, for this particular decision, there was no difference, or no significant advantage among the alternatives on the air quality factor. Paz was able to say to this nurse that we know air quality is important. However, because the team has already taken care of that concern and all the alternatives have performed well regarding air rate exchange, that's not the most important advantage of this decision.

What we could see from the relevant facts, meaning the attributes of the alternatives, was that the need for maintenance was really different from system to system. When we asked the facility managers, the people who take care of the building once it's working, what it would mean to perform those maintenance tasks, they explained to us that they would need to close certain areas of the children's hospital more frequently with one of the mechanical systems. That would interrupt the hospital's operations, and because the different systems had very different requirements in terms of maintenance cadence, that difference was the most important advantage of this decision.

So, to conclude the example: (1) the statement "air quality is the most important factor when choosing HVAC systems at hospitals" is an assessment/opinion, and (2) in this context, all alternatives performed well, judging by their attributes (assertion/factual statement). Therefore, the difference between alternatives was not relevant to the decision. However, this does not mean that air quality should be forgotten when designing the wider system, but it was not an important advantage in this decision.

Suhr (1999) states the following principle: *Decisions must be based on the relevant facts.* He calls this the principle of anchoring. This is not to be confused with anchoring bias, first described by Kahneman and Tversky (Kahneman, 2011), when people mix information that is not related. When Suhr talks about anchoring, it is about describing advantages against the same starting point, which is the least preferred attribute in one of the alternatives, our baseline, according to the criteria the team has agreed on. Therefore, the alternative with the least preferred attribute will not have an advantage within this particular factor.

When describing advantages, we always use the same starting point. Going back to the basketball example, if taller is better, we will compare Player 1 on the left in Figure 4.4 against all the players on his right (Players 2 and 3), since Player 1 has the least preferred attribute. Also, we do not assume that every difference in height is equally important; remember, do not assume linearity.

The anchoring principle builds on the previous one: *decisions must be based on the importance of advantages (IoA).* To understand an advantage well, you

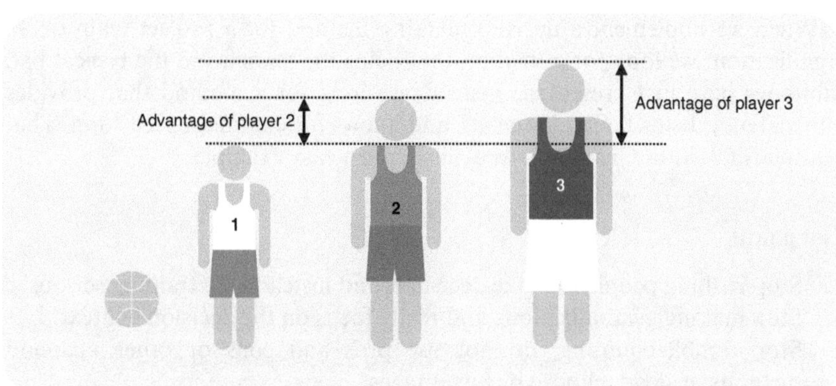

Figure 4.4 Comparing heights with more than two alternatives

need to know the relevant facts; in this case, you need to know the advantages in height for each player and who you are comparing against.

What do we want in a decision-making method?

Not all decision-making methods are the same, and it is important to identify characteristics that make a method viable for making sound, transparent, and collaborative decisions that help decision-makers avoid bad habits. What is a good decision-making method? In summary, we want three characteristics (Arroyo, 2014):

1. *Providing transparency*: A good decision-making method helps decision-makers organize information transparently. This means that it helps create a clear and shared rationale for a decision. As mentioned in Chapter 3, we want to compare the value and cost of the alternatives separately.
2. *Building consensus*: Guides conversations to reveal concerns from all stakeholders, helps decision-makers to anchor their assessments to the context of the decision, prevents double-counting information, and helps in reaching consensus while avoiding conflicts created when we try to agree on generalizations. Notice here that consensus does not mean that everyone thinks the same, but that people making the decision understand the why and commit to moving in the same direction to act upon the decision.
3. *Continuous learning*: Helps decision-makers to explain and document the decision rationale easily. This helps save time in future iterations when you learn something new about an alternative or a new perspective.

Summary

We call them bad habits because they are hard to change. We do not expect miracles, but we know they can be changed with practice, patience, and the right decision-making methods and processes; the reward is better decisions.

When we implement a decision-making method for a project team or an organization, we look for a method that helps the team avoid the typical bad habits we have just presented. Instead, we look for a method that provides transparency, helps build consensus, and allows for continuous learning. That method is CBA! In Chapter 5, we will share an easy example.

Key points

- Stop rushing people to make decisions and instead take training seriously.
- Stop making generalizations and really focus on the decision context.
- Stop double-counting; do not use pros and cons or other unsound methods; instead, choose by advantages.
- Stop assuming linear trade-offs; instead, build preference curves that represent the context of the decision.

- Stop confusing opinions for facts; instead, stay curious and always refer back to relevant facts when making a decision. Consider which differences between the attributes of the alternatives are relevant.

References

Arroyo, P. (2014). "Exploring Decision-Making Methods for Sustainable Design in Commercial Buildings." Doctoral dissertation, University of California, Berkeley.

Arroyo, P., Christensen, R., Schöttle, A., and Long, D. (2019). "Lessons Learned on Teaching Choosing by Advantages." In *Proceedings of the 27th Annual Conference of the International Group for Lean Construction (IGLC)*, pp. 405–416.

Austin, J.L., (1975). *How to Do Things with Words*. Oxford: Oxford University Press.

Dreyfus, S.E., and Dreyfus, H.L. (1980). "A Five-Stage Model of the Mental Activities Involved in Directed Skill Acquisition." Operations Research Center, University of California, Berkeley.

Flores, F. (2012). *Conversations for Action and Collected Essays: Instilling a Culture of Commitment in Working Relationships*. CreateSpace Independent Publishing Platform.

Flores, G.P. (2016). *Learning to Learn and the Navigation of Moods: The Meta-Skill for the Acquisition of Skills*. Pluralistic Networks Publishing.

Kahneman, D. (2011). *Thinking, Fast and Slow*. London: Penguin Random House.

Parrish, K., and Tommelein, I.D. (2009). "Making Design Decisions Using Choosing by Advantages." In *Proceedings of the 17th Annual Conference of the International Group for Lean Construction*, pp. 501–510.

Saaty, T.L. (1988). *What Is the Analytic Hierarchy Process?* Berlin: Springer.

Searle, J.R. (1969). *Speech Acts: An Essay in the Philosophy of Language* (vol. 626). Cambridge: Cambridge University Press.

Spetzler, C., Winter, H., and Meyer, J. (2016). *Decision Quality: Value Creation from Better Business Decisions*. Hoboken, NJ: John Wiley & Sons, Inc.

Suhr, J. (1999). *The Choosing By Advantages Decisionmaking System*. Westport, CT: Quorum.

5 CBA

A sound decision-making method

As stated before, we want you to use sound decision-making methods. We have already explained the main principles of Choosing By Advantages (CBA) in Chapter 4, and, yes, it can be difficult to grasp a method fully in theory. To make these concepts more tangible, we will present a simple example here and explain different CBA methods step-by-step.

CBA methods

Suhr (1999) points out that different types of decisions call for different types of methods, and, in practice, we have seen this principle add value. If you need to decide between two alternatives, you should probably use a simpler method than when choosing among three or more alternatives. The complexity of the decision will determine what type of method to use. There is no need to overcomplicate things; this will not make the decision more transparent or sound.

Suhr developed a suite of CBA methods for different decision-making problems. All CBA methods build on the key principles described in Chapter 4. Remember, the fundamental rule is that *decisions must be based on the importance of advantages*, not on the importance of dollars. The following are the CBA methods, presented from simpler to more complex:

Starting with simple methods for very simple decisions:

1. Simplified Two-List method
2. Instant CBA
3. Recognition Response CBA

For more complex decisions:

4. Two-List method (two alternatives)
5. Tabular method (two or more alternatives)

DOI: 10.1201/9781003518310-6

To start, we will explain the Tabular method, which is the more structured method to learn. Following the Tabular method step-by-step will help you see the difference from what you might normally do. We recommend using this method if you have more than two alternatives. We also recommend the Tabular method if you have two alternatives and you are new to CBA. We want to ensure you follow the underlying principles before using the simpler methods. That is also why, in our workshops, we always teach the Tabular method first. We will explain the CBA Tabular method using a step-by-step guideline and an easy example with only two alternatives, just for simplicity. This will help you consolidate your understanding of the principles presented in Chapter 4; after you have learned this method, you can easily understand the simpler methods.

CBA Tabular method

Here, we will use a simple example to demonstrate CBA steps. Imagine you have to choose a new phone.[1] Figure 5.1 shows the seven steps required to implement the CBA Tabular method.

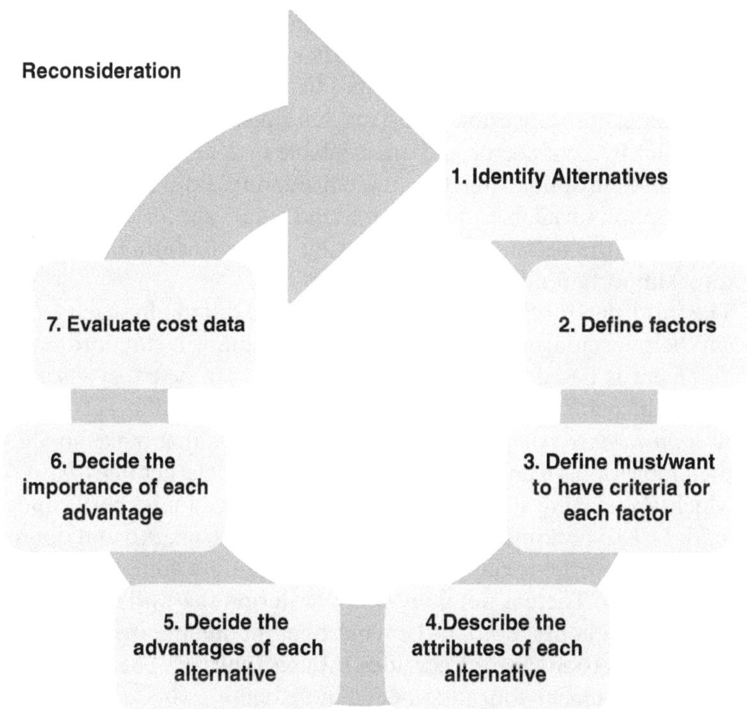

Figure 5.1 Steps to implement the CBA Tabular method

In order to apply the CBA method, the stages are as follows:

1 Identify the alternatives to be considered in the decision-making process. In this case, we will decide between an iPhone 15 and an iPhone 15 Pro Max. We want to keep this example simple and relatable, so we decided not to consider other iPhone models. We are also assuming that the decision-maker wants to stay in the Apple ecosystem, and therefore, we are not evaluating against other major smartphones, such as Samsung Galaxy. Just to be clear: we are not recommending any particular smartphone, because, of course, it depends on your personal preferences. However, we will walk you through our rationale in making this decision for one person, whom we will call Susan, so you can make your own judgment later on; but, more importantly, you can check that you are asking the right questions and comparing the right information.
2 Susan's second step will be to define factors that differentiate the alternatives. Please note that cost is not included as a factor in CBA because it is treated as a constraint on the decision, and it will be incorporated in the final step of this method. After the initial review on the Apple webpage (Apple, 2023), Susan realized that both models differ in screen size, weight, camera, chip technology, and battery life. She understood that several factors could be ignored in this decision since they have equal attributes for both models, and, therefore, no alternative has an advantage. For example, both iPhones have the same pixel density (460 ppi), have 5G, Face ID, and the same water-resistance and audio playback attributes. Both iPhones have a single USB-C connector and are available in 256GB and 512GB options, which are the options that Susan is considering. Both iPhones have several color options available, and Susan found colors she liked for both models. Therefore, the decision is simplified by just considering the factors that reveal important differences for her.
3 The third step for Susan will be to decide the criteria for each factor. This will help her judge the alternatives. A criterion is a standard on which a judgment is based (Suhr, 1999). The criteria can be either *must-have* criteria, which are standards that an alternative must fulfill to be considered, or *want-to-have* criteria, which are the standards that make an alternative preferable or have an advantage. Susan first defines her *must-have* criteria, which are staying in communication with her family, performing some work tasks, checking emails, taking pictures, listening to audiobooks, and using her favorite social media apps. Both phones comply with her *must-have* criteria. There is not a *must-have* criterion that will make her discard one of the iPhones; both of them meet her minimum expectations. To simplify the decision, we will not add them to Table 5.1, but you could if you want to document your rationale in more detail.

   Susan realizes that she has several *want-to-have criteria*. When it comes to screen size, her preference is to have a smaller screen; 15.49 cm/6.1" (diagonally) is already big for her hands; note that another user may find bigger is better for screen size, especially if they plan to watch videos on

Table 5.1 Factors and criteria

Factor	*Criterion*
Screen size	Smaller is better
Weight	Less is better
Chip	Latest technology is better
Camera	More optical zoom is better
Battery life	Longer is better

their phone or want to use a bigger font size. In terms of the iPhone's weight, less is better for her. In terms of chips, the latest technology is better, so the apps she uses respond quickly. However, she is not a video game player, so she is not big on graphic performance. In terms of the phone's camera, she notices the differences in optical zoom availability, and she knows that the more optical zoom, the better for her. Finally, for battery life, longer is better for her. Table 5.1 summarizes the factors and criteria for Susan.

4 The fourth step for Susan is to summarize the attributes of the alternatives. Remember that attributes are characteristics of the alternatives; they can be qualitative or quantitative, but they are not judgmental statements. They should be independent of the other alternatives; it is not a comparison. To help with the decision, Susan also underlines the least preferred attribute in each factor according to the factor's criteria (see Table 5.2).

5 Susan's fifth step is to decide each alternative's advantages. In this step, we look at the alternatives' attributes, and we always compare them against the least preferred attribute based on the defined criterion for each factor. For example, for screen size, the advantage is 1.53 cm/0.6" smaller for iPhone 15. On the other hand, notice that the alternative with the least preferred attribute, meaning the worst alternative for that factor, does not have an advantage. The least preferred attribute is the anchoring point or baseline, and the advantage cell will be left empty for that alternative. For example, the iPhone 15 Pro Max does not have an advantage regarding screen size, so both the advantage and the importance of advantage cells will be left empty. In summary, the iPhone 15 has the advantage of being 1.53 cm/0.6" smaller and weighing 50 g less (1.76 oz). On the other hand, the iPhone 15 Pro Max has the advantage of having newer chip technology, 5x versus 2x optical zoom, and 9 hours more of video playback for battery life. Table 5.3 shows the advantages of each alternative.

6 Susan's sixth step is deciding the importance of the advantages (IoA). The IofA corresponds to a value, in the sense of worth, given to the alternatives based on the relevance of its advantages. To do that, Susan looks at the magnitude of the advantages and the context of the decision; she tries to understand

Table 5.2 CBA Tabular method: Step 4 Describe attributes

Factor (Criterion)	*iPhone 15 attributes*		*iPhone 15 Pro Max attributes*	
Screen size (Smaller is better)	Att.: 15.49 cm (6.1") diagonal screen size		Att.: 17.02 cm (6.7") diagonal screen size	
	Adv.:	IoA:	Adv.:	IoA:
Weight (Less is better)	Att.: 171 grams (6.02 ounces)		Att.: 221 grams (7.81 ounces)	
	Adv.:	IoA:	Adv.:	IoA:
Chip (Latest technology is better)	Att.: A16 Bionic chip		Att.: A17 Pro chip	
	Adv.:	IoA:	Adv.:	IoA:
Camera (More optical zoom is better)	Att.: Advanced dual-camera system, zoom options: 0.5x, 1x, 2x		Att.: Pro camera system, zoom options: 0.5x, 1x, 2x, 5x	
	Adv.:	IoA:	Adv.:	IoA:
Battery life (Longer is better)	Att.: 20 hours of video playback		Att.: 29 hours of video playback	
	Adv.:	IoA:	Adv.:	IoA:

Notes: Underlining indicates the least preferred attribute.
Adv. = advantage; Att. = attribute; IoA = Importance of Advantages.

the impact of those advantages in her use of the iPhone. The sums of the IofAs for all alternatives represent the total importance of the advantages of each alternative. To assess the advantages, Susan follows these steps:

a. Identify the most important advantage per each factor (in bold in Table 5.3).
b. Select the "paramount advantage," which is the advantage with the greatest importance, and assign the highest IofA score to it. In this case, she assigns a 100 IofA to 5x optical zoom vs. 2x; when looking at this difference capability from 2x to 5x, she realizes that this will make a big difference in having great pictures from a distance at her fingertips, this will be very useful for outdoor pictures since she likes to hike and capture birds or other animals.

Note: the scale of importance can also consist of different ranges, such as 0–10 or 0–200. In our experience, using a scale of 0–100 is easy to apply and gives you enough room to assess the importance of multiple advantages.

c. Assess the most important advantage of each factor compared to the paramount advantage. Once she knows that 5x optical zoom vs. 2x is the paramount advantage, she can use that to compare it against the rest of the advantages. See Table 5.2, where Susan wrote down her scale of importance; the advantage of 1.53 cm (0.6") smaller

Table 5.3 CBA tabular method: Step 5 Decide on the advantages

Factor (Criterion)	*iPhone 15 attributes*		*iPhone 15 Pro Max attributes*	
Screen size (Smaller is better*)	Att.: 15.49 cm (6.1") diagonal screen size		Att.: 17.02 cm (6.7") diagonal screen size	
	Adv.: 1.53 cm (0.6") smaller	IoA:	Adv.:	IoA:
Weigh (Less is better)	Att.: 171 grams (6.02 ounces)		Att.: 221 grams (7.81 ounces)	
	Adv.: 50 g (1.76 oz) less	IoA:	Adv.:	IoA:
Chip (Latest technology is better)	Att.: A16 Bionic chip		Att.: A17 Pro chip	
	Adv.:	IoA:	**Adv.: newer chip technology**	IoA:
Camera (More optical zoom is better)	Att.: Advanced dual-camera system, zoom options: 0.5x, 1x, 2x		Att.: Pro camera system, zoom options: 0.5x, 1x, 2x, 5x	
	Adv.:	IoA:	**Adv.: 5x optical zoom vs. 2x.**	IoA:
Battery life (Longer is better)	Att.: 20 hours of video playback		Att.: 29 hours of video playback	
	Adv.:	IoA:	**Adv.: 9 hours more of video playback**	IoA:

Notes: * To state qualitative advantages, you can use, for example, somewhat more, more, much more, very much more, to make the comparisons clear.

Bold indicates the most important advantage in every factor; underlining indicates the least preferred attribute.
Adv. = advantage; Att. = attribute; IoA = Importance of Advantages.

screen size is relevant but not so much compared to the difference in the zoom, so she decided on a score of 80 IoA, then 9 hours more of video playback battery is also desirable, but not as much as the other two advantages, since she will probably charge her battery every night anyway; however, those 9 hours will probably make a difference in long business trips, she decides on a score of 60 IoA. Finally, on the newer chip technology, she decides on 30 IoA. Finally, for the 50 grams less, 20 IoA since she will probably not notice those differences a lot. Table 5.4 shows the advantages on the scale of importance.

d. Note that if you have more than two alternatives, you need to decide on the importance of the remaining advantages. We will explain this in detail in Chapter 6.

Table 5.4 Advantages rated on the scale of importance

Importance	*Advantage*
100	5x optical zoom vs. 2x
90	
80	1.53 cm (0.6") smaller
70	
60	9 hours more of video playback
50	
40	
30	newer chip technology
20	50 g less (1.76 oz)
10	
0	

After all the advantages are weighted, Susan can add up the total IoA for each alternative. In this case, the iPhone 15 has 100 total IoA, and the iPhone 15 Pro Max has 190 total IoA. Table 5.5 summarizes the CBA Tabular method's steps 1–6.

7 Finally, in step seven, Susan has to evaluate cost data vs. IofAs. If the alternatives have the same cost, the decision is easy; she should choose the alternative with the higher IofAs. In this case, the alternatives have a different price:
   - The price for 256GB iPhone 15 is $899.
   - The price for 256GB iPhone 15 Pro Max is $1199.

The next section of this chapter will discuss how to make the final decision. The main question here is how to correctly use cost information. In subsequent sections, we will introduce the remaining CBA methods.

How do you correctly use cost information?

As stated before, we will consider differences in cost as a constraint, not as a means to evaluate the value of an alternative. You must use cost information correctly when you have alternatives with different costs or unequal money decisions, as Jim Suhr called them. As discussed in Chapter 3, we tend to hyper-focus on costs and do a poor job of understanding value. However, now that you understand the advantages, you have a better sense of what value is. In this case, Suhr (1999) recommended developing a chart of total importance vs. cost (such as in Figure 5.2). However, we want to warn people not to use this chart to develop ratios and create mindless rules. Remember, the whole purpose of this method is to stay in touch with the decision context and the differences between the alternatives, not to create general rules and then use cost data to make decisions. If you only want to decide based on cost and select the cheapest alternative, you can still do this, but then you will not be implementing CBA, and you are probably not getting the most value out of your money.

Table 5.5 CBA Tabular method's steps 1–6

Factor (Criterion)	*iPhone 15 attributes*		*iPhone 15 Pro Max attributes*	
Screen size (Smaller is better*)	Att.: 15.49 cm (6.1") diagonal screen size		Att.: 17.02 cm (6.7") diagonal screen size	
	Adv.: 1.53 cm (0.6") smaller	IoA: 80	Adv.:	IoA:
Weigh (Less is better)	Att.: 171 grams (6.02 ounces)		Att.: 221 grams (7.81 ounces)	
	Adv.: 50 g (1.76 oz) less	IoA: 20	Adv.:	IoA:
Chip (Latest technology is better)	Att.: A16 Bionic chip		Att.: A17 Pro chip	
	Adv.:	IoA:	**Adv.: newer chip technology**	IoA: 30
Camera (More optical zoom is better)	Att.: Advanced dual-camera system, zoom options: 0.5x, 1x, 2x		Att.: Pro camera system, zoom options: 0.5x, 1x, 2x, 5x	
	Adv.:	IoA	**Adv.: 5x optical zoom vs. 2x**	IoA: 100
Battery life (Longer is better)	Att.: 20 hours of video playback		Att.: 29 hours of video playback	
	Adv.:	IoA	**Adv.: 9 hours more of video playback**	IoA : 60
Total importance		100		190

Notes: * To state qualitative advantages, you can use, for example, somewhat more, more, much more, very much more, to make the comparisons clear.

Bold indicates the most important advantage in every factor; underlining indicates the least preferred attribute.

Adv. = advantage; Att. = attribute; IoA = Importance of Advantages.

Going back to Susan's iPhone example, she makes a graph to analyze the total importance of each alternative's advantages versus the cost (or price) to make the decision; this will help her ask the right questions (Figure 5.2).

Susan now has to decide: Do I want to spend $300 more to obtain an alternative with 190 instead of 100 importance of advantages? Are those 90 IoA more worth $300 for me? This decision is personal, and it has to be viewed holistically, considering other decisions that she is making. What else can she get with those $300? Does she have the money available for the iPhone 15 Pro Max? She can also think outside of the box and check out the other alternatives, such as the iPhone 15 Pro and the iPhone 15 Plus. That would be part of the reconsideration phase and should be considered formally in the decision-making process.

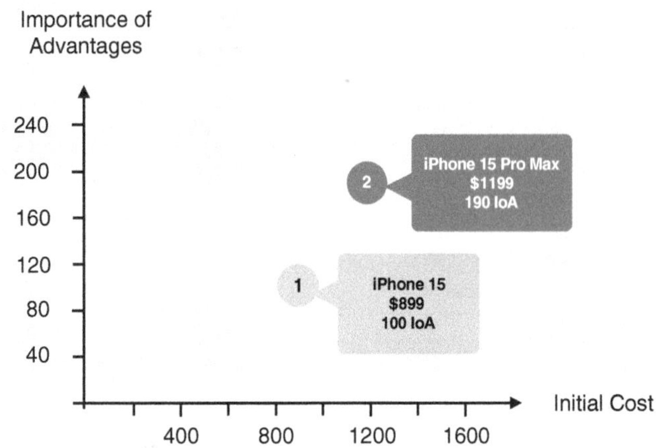

Figure 5.2 Importance of advantages vs. price chart

Both decisions will be reasonable in this case. This is a hard choice since the most valuable alternative is also more expensive; however, that is not always the case, as we discussed in Chapter 3. When implementing CBA, she will have clarity of what value she is getting for the money she is spending. There is much more to explore in terms of considering costs when making project decisions, including differentiating between initial and life-cycle costs. However, we will discuss that in the following chapters.

Two-List method

Do you think the CBA Tabular method is too hard to remember? Definitely, yes, if this is your first time. However, it becomes easier and easier as you practice. Suhr (1999) stated that you can simplify the method for simple decisions by taking fewer steps. The Two-List method is a simplification of the Tabular method. You should use the Two-List method for simple decisions involving only two mutually exclusive alternatives. Here are the steps:

1. Mentally perceive the attributes of each alternative one factor at a time and simultaneously decide the least preferred attribute.
2. Write each advantage (compared to the least preferred attribute) in a list beneath each alternative. This could be on Post-its on a wall or similar.
3. Decide the importance of each advantage by first selecting the paramount advantage to establish a scale of importance. Assess the importance of the

Table 5.6 Two list method

Advantages of iPhone 15		*Advantages of iPhone 15 Pro Max*	
Adv.: 1.53 cm (0.6") smaller	IoA: 80	Adv.: newer chip technology	IoA: 30
Adv.: weighs 50 g (1.76 oz) less	IoA: 20	**Adv.: 5x optical zoom vs. 2x**	IoA: 100
		Adv.: 9 hours more of video playback	IoA: 60
Total importance	100	Total importance	190

Note: bold indicates most important attribute.
Adv. = advantage; IoA = Importance of Advantage.

advantages on the same scale. This can also be done by moving the Post-its up and down between the advantage you are assessing and the paramount advantage.

4. Choose the alternative with the greatest total importance (sum) of advantages if there is no cost difference, or do an IoA vs. cost chart if you have a cost difference between the alternatives.

As you can see in Table 5.6, Susan could have directly listed the advantages of each iPhone model and then written the importance of each advantage.

Simplified Two-List method

Are you ready for further simplification? This is the Simplified Two-List method for simple decisions involving only two mutually exclusive alternatives. This method is easy to use when alternatives are of equal cost. We know the alternatives here are not the same price, but we will use the same example for simplicity. Since no total importance of advantage is used here, it gets harder to evaluate differences in cost.

1. Mentally consider the attributes of each alternative, one factor at a time, simultaneously deciding the least preferred attribute.
2. Write each advantage (compared to the least preferred attribute) in a list beneath each alternative.
3. Without deciding the importance of each advantage individually, choose the alternative having the more important set of advantages.

As you can see in Table 5.7, Susan could have directly listed the advantages of each iPhone model and then written the importance of the set of advantages for each alternative.

Table 5.7 Simplified two list method

Advantages of iPhone 15	*Advantages of iPhone 15 Pro Max*
1.53 cm (0.6”) smaller	Newer chip technology
Weighs 50 g (1.76 oz) less	5x optical zoom vs. 2x
	9 hours more of video playback

Instant CBA

You can even further simplify using the Instant CBA method for simple decisions involving a choice between two mutually exclusive alternatives. It is easy to use when alternatives are of equal cost. Otherwise, you also have to mentally evaluate costs.

- Mentally form a clear, accurate, and sensory-rich perception of the advantages of the alternatives, and at the same time, choose the alternative having the more important set of advantages.

Mentally, Susan understands and has a sensory-rich perception that the advantages of the iPhone 15 Pro Max (newer chip technology, 5x optical zoom vs. 2x, and 9 hours more of video playback) are more important than the advantages of the iPhone 15 (0.6” smaller and weights 50 g (1.76 oz) less. In this case, you do not need a table!

Recognition response CBA

Finally, the most simplified version would be the Recognition response CBA for simple decisions involving a choice between two mutually exclusive alternatives.

- Observe, recognize the situation, and respond with a decision.

If the basis of the decision considers information from previous experiences using CBA, one can use Recognition response CBA.

Situations where the Recognition response CBA would be appropriate are when you have made this decision before. An example would be if Susan just got a new iPhone and lost it; she already knows what she values more, so she can just order a replacement, assuming no new versions are available, and all information stays the same as her last decision.

Summary

To use CBA in a complex decision, you should follow this step-by-step process. In Chapter 6, we will provide a more complex example where a team is making a decision, and we will also show you how to deal with three or more alternatives.

Key points

- CBA has several decision-making methods for different situations. However, all CBA methods make decisions based on the importance of advantages.
- For more than two alternatives, use the Tabular method.
- If you are new to CBA, we recommend you start with the Tabular method. You will need more clarity, and writing down the attributes, factors, and criteria will help you get used to the vocabulary before jumping to writing advantages.
- Cost is evaluated at the end, and you can make a chart to compare the total importance of advantages vs. cost.

Note

1 This was Paz's decision in Chapter 3 but updated to the year 2023.

References

Apple. (2023). https://www.apple.com/iphone/compare/ (accessed January 12, 2023).

Suhr, J. (1999). *The Choosing By Advantages Decisionmaking System*. Westport, CT: Quorum.

6 The CBA Tabular method

Sustainable museum case study

When studying teams that have produced sustainable construction projects, we have noticed that they organized their decisions in a different way. They are able to make decisions on time and with a long-term mindset, so, during the design phase, they made the time to hear about different perspectives and understand value. This is typically an effort for the teams. They invest time upfront to generate data and better understand and align the need and purpose of the decisions.

In this chapter, we provide a case study to explain how to use Choosing By Advantages (CBA) in a more technical decision. We will demonstrate the use of CBA to choose a mechanical system for a museum building in San Francisco, California. This case study is based on a research paper developed by Arroyo et al. (2016). We present a decision considering three alternatives. Therefore, the CBA Tabular method is used (Suhr, 1999). As stated before, when using CBA, we help the team make decisions transparently, build consensus, and allow for documentation and continuous improvement.

Case study background

There were many more perspectives when it came to defining value, and that was evident in the interviews. The owner, the architect, the mechanical engineer, and the facility manager all had different perspectives. The team was encouraged to look at value considering multiple factors. The facility manager was concerned about how reliable and easy to maintain each of the systems was; the owner was concerned with the water consumption and the possibility of achieving a Net Zero Energy (NZE) building, a building that produces the same energy that it uses. The architect was concerned with the amount of space the system took from the museum exhibits and the noise it would make, as it affected the user experience. The mechanical engineer was concerned with achieving NZE and reducing CO_2 emissions during operations. The building also required a significant retrofit of an old pier in San Francisco, where the museum was being relocated. Finally, the owners were concerned with managing the project budget and schedule in a way that would ensure success and

DOI: 10.1201/9781003518310-7

show the donors that their resources had been put to good use. From the beginning, the team wanted the building to show how sustainable design and construction could be achieved (Woolsey, 2012). The following were the design goals and building features they wanted.

Design goals and building features

The new museum campus size is 30,658 m^2 (330,000 ft^2), including indoor and outdoor exhibit space and 0.6 ha (1.5 acres) of freely accessible public space. The project has several goals that were incorporated into the design as a result of the evaluation of design alternatives:

1. *Net zero energy (NZE)*: The design team targeted NZE. Therefore, the building produces as much energy as it consumes on site in a year.
2. *Water usage goal*: The design team wanted to minimize water usage.
3. *Educating about energy-efficient design*: The design team wanted to educate visitors about energy-efficient design, and the museum itself was designed to serve as an interactive exhibit.

Step-by-step CBA Tabular method

As mentioned, we are illustrating the application of the CBA Tabular method to choose a heating, ventilation, and air conditioning (HVAC) system for this interactive museum in San Francisco. Here, we share the decision-making results for each step.

Step 1: Identify alternatives

Choosing an HVAC system involves many decisions (Lechner, 2008). The design team needed to choose the primary energy source for heating systems (e.g., gas or electricity), the distribution system (e.g., radiant slab or fan coil unit), and the heat sink for pumping heat from the building (e.g., the outdoor air or a body of water).

This case study analyzes three different but interdependent decisions: (1) the energy source for the heating system, (2) the energy source for the cooling system, and (3) the sink to pump heat from the building (bay water or air using a cooling tower). Table 6.1 shows the alternatives that were considered by the design team.

All options include roof insulation, energy-efficient lighting, and basic daylighting controls. In addition, the mechanical engineer recommended the use of a radiant floor system as a distribution system for all alternatives (a radiant floor system, consisting of pipes running through the floor that are embedded in the floor of the building, to move water at a desired temperature to condition the space).

Table 6.1 Design alternatives for energy sources and heat sinks

	Alternative 1	*Alternative 2*	*Alternative 3*
Heating source	Natural gas boilers	Bay water coupled with electric heat pumps sourced by the utility company	Bay water coupled with electric heat pumps with PV panels
Cooling source	Electric chiller source by utility company	Bay water coupled with electric heat pumps sourced by the utility company	Bay water coupled with electric heat pumps with PV panels
Heat rejection	Cooling tower	Bay water	Bay water

The three alternatives considered in this test case are:

- *Alternative 1*: Natural gas boiler and chiller system with a cooling tower. This alternative is commonly used in commercial buildings and is considered the standard way of designing HVAC systems. In this case, the heating source is a natural gas boiler, and the cooling source is an electric chiller. Heat rejection is done through a cooling tower, which evaporates water.
- *Alternative 2*: Electric heat pump sourced by utility company with bay water. This alternative takes advantage of the building's position to use bay water for cooling and heating. The idea is to exchange heat between the cooler bay water and the hot water coming from the building after the bay water is released. In this case, the heating and cooling source will be bay water coupled with electric heat pumps connected to a local utility company. Heat rejection is accomplished through bay water recirculation.
- *Alternative 3*: Electric heat pump sourced from a photovoltaic (PV, solar panels) system with bay water. In this case, the heating and cooling source will be bay water coupled with electric heat pumps (the same as alternative 2) connected with a PV system as the energy source instead of the local utility company. Heat rejection is accomplished through bay water recirculation (the same as alternative 2).

Step 2: Define the factors

In CBA, the design team needs to identify factors that will help differentiate between alternatives. The process of identifying factors may be iterative, since the design team may find new information when they develop a more detailed design of the alternatives. Based on the available data, seven factors help the team to differentiate between the alternatives. The factors represent the views of the owner, the architect, the mechanical engineer, the operation manager, and the users. The factors and the rationale for selecting them are as follows:

1. *Experience in using HVAC systems*: The alternatives have important differences in how often they are utilized. The majority of commercial buildings use cooling towers and water-based systems, which are infrequently utilized. It is important to have a proven and reliable system.
2. *Space requirements*: The alternatives have important differences in space requirements due to the requirement of having (or not) a cooling tower. Space is scarce and can be used for museum displays instead of a cooling tower.
3. *Contribution to the goal of NZE*: The alternatives differ in how they contribute to the goal of NZE. For example, using a PV panel system allows for onsite generation vs. natural gas that needs to be sourced elsewhere.
4. *Water usage*: The alternatives differ in their requirements for municipal fresh water. The cooling tower recirculates fresh water and loses significant quantities in evaporation, whereas the bay water system does not require precious fresh water from public sources.
5. *Maintainability*: The alternatives differ in terms of maintenance requirements. For example, using a bay water system poses special maintenance challenges compared to standard cooling tower maintenance procedures.
6. *CO_2 emissions*: The alternatives differ significantly in terms of CO_2 emissions during the operation and maintenance phase.
7. *Noise*: The alternatives differ in terms of noise due to airflow movement. For example, a cooling tower emits noise compared to a bay water system, which does not require a cooling tower.

Step 3: Define the 'must' and 'want to have' criteria for each factor

The design team needs to agree on the criteria for each factor to judge the alternatives. Some attributes have a standard evaluation, in which case it is easy to establish a criterion (e.g., for CO_2 emissions, the criterion may be the lower the CO_2 emissions, the better). In other cases, the design team needs to describe what they want (e.g., for experience using the HVAC systems, the criterion may be, the more reliable the system is, the better). The criteria do not represent a trade-off. CBA does not assume that every increment of performance is equally valuable. Table 6.2 presents the factors and criteria considered in this decision in the first column.

Step 4: Summarize the attributes of each alternative

We used data available from the mechanical engineers to obtain the attributes of each alternative. Table 6.2 summarizes the attributes of the alternatives. The

Table 6.2 The CBA Tabular method

Factor (Criterion)	*Alternative 1: Natural gas boiler and chiller system with a cooling tower*		*Alternative 2: Electric heat pump sourced by utility company with bay water*		*Alternative 3: Electric heat pump sourced by PV panel system with bay water*	
Experience using this HVAC system	Att.: Typically used in commercial buildings. This is the standard way.		Att.: It is not that common to use bay water for HVAC systems; unanticipated problems may arise.		Att.: It is not that common to use bay water for HVAC systems; unanticipated problems may arise.	
(The more reliable the system is, the better.)	**Adv.: It is more reliable than alternatives 2 and 3.**	IoA: 50	Adv.:	IoA:	Adv.:	IoA:
Space requirements	Att.: The cooling tower uses a lot of space. 35 m² (370 ft²) approximately.		Att.: The condenser does not occupy outdoor space.		Att.: The condenser does not occupy outdoor space.	
(The less space the HVAC system uses, the better.)	Adv.:	IoA:	**Adv.: It saves around 35 m²**	IoA: 20	**Adv.: It saves around 35 m²**	IoA: 20
Contribution to goal of NZE	Att.: It requires external energy by using natural gas and electricity from the grid. It does not allow for NZE.		Att.: Lower power consumption than air-cooled systems, especially at peak load. It requires external energy by using natural electricity from the grid. It does not allow for NZE.		Att.: Lower power consumption than air-cooled systems, especially at peak load. It allows the building to produce the same energy that it consumes.	
(The more the alternative contributes to achieve the NZE target, the better.)	Adv.:	IoA:	Adv.: Slightly better than alternative 1.	IoA: 20	**Adv.: Significantly better than alternative 1, allowing for NZE.**	IoA: 100

Water usage	Att.: It requires the use of evaporative cooling towers, which use an estimated 7.57 million liters (2 million gallons) of fresh water per year.		Att.: Used water is returned to the bay.		Att.: Used water is returned to the bay.	
(The less water the system uses, the better.)	Adv.:	IoA:	**Adv.: It saves 7.57 million liters of fresh water per year.**	IoA: 35	**Adv.: It saves 7.57 million liters of fresh water per year.**	IoA: 35
Maintainability	Att.: Easy standard maintenance.		Att.: Hard maintenance. biofouling will be produced.		Att.: Hard maintenance. biofouling will be produced.	
(The easier to maintain, the better.)	**Adv.: Maintenance is easier and less frequent than alt 2 and 3.**	IoA: 60	Adv.:	IoA:	Adv.:	IoA:
CO_2 emissions	Att.: **260,698 kg (574,740 lb) CO_2 per year.**		Att.: 164,744 kg (363,198 lb.) CO_2 per year.		Att.: No operation CO_2 emissions per year.	
(The lower the CO_2 emissions, the better.)	Adv.:	IoA:	Adv.: Avoids 95,954 kg CO_2 per year.	IoA: 40	**Adv.: Avoids 260,698 kg CO_2 per year.**	IoA: 80
Noise	Att.: It may produce noise problems associated with the high airflow required for air-cooled systems.		Att.: It is a quiet system.		Att.: It is a quiet system.	
(The less noise, the better.)	Adv.:	IoA:	**Adv.: Less noisy than alternative 1.**	IoA: 10	**Adv.: Less noisy than alternative 1.**	IoA: 10
Total IoAs		**110**		**125**		**245**

Notes: Att. = attribute; Adv. = advantage. IoA = Importance of Advantages. Underlining indicates the least preferred attribute; bold indicates the most important advantage in each factor.

least preferred attributes are underlined and will be used as comparison points to describe advantages.

Step 5: Decide the advantages of each alternative

Once the attributes have been summarized, the design team needs to apply the criteria to identify the advantages. Table 6.2 presents the advantages of each alternative for each factor.

Step 6: Decide the importance of each advantage

This part of the process is collaborative, and decisions are reached through discussion within the design team. The most important advantage for each factor is shown in bold in Table 6.2. The rationale for choosing the IoA scores is explained here:

- You can see that the paramount advantage or most important advantage, which is assigned a value of 100, is that alternative 3 is significantly better than alternative 1 in terms of contributing to the NZE goal. Actually, only alternative 3 allows for achieving an NZE building. This is one of the main purposes of the building and also allows for teaching sustainability concepts to visitors. This advantage is used as a point of comparison with the other advantages of the more important advantages in all other factors (advantages in bold in Table 6.2). In addition, alternative 2 is slightly better than alternative 1 with regards to contributing to NZE because it does not require natural gas. However, that seemed less important when compared to the paramount advantage, so it has an IoA of 20.
- Alternative 3 avoids 260,698 kg (574,740 lb) of CO_2 per year when compared to alternative 1. This seems important, so an IoA of 80 was assigned. At the same time, alternative 2 avoids 95,954 kg (211,542 lb) of CO_2 per year when compared to alternative 1, so an IoA of 40 was assigned.
- Alternative 1 has two advantages over alternative 2 and 3. First, it is more reliable than alternatives 2 and 3, which seems important, but not as much as the paramount advantage. Accordingly, it has been assigned 50 IoA. Second, alternative 1 maintenance is easier and less frequent than alternatives 2 and 3, which is also important, so an IoA of 60 was assigned.
- Lastly, alternatives 2 and 3 have three equal advantages over alternative 1. First, alternatives 2 and 3 save approximately 35 m^2 (370 ft^2) when compared with alternative 1, because they do not require a cooling tower. This advantage seems relatively unimportant when compared with the paramount advantage; therefore, an IoA of 20 was assigned. Second, alternatives 2 and 3 save 7.57 million liters (2 million gallons) of fresh water per year when compared to alternative 1. Compared to the paramount advantage, this seems somewhat important, so an IoA of 35 was

assigned. Third, alternatives 2 and 3 are less noisy than alternative 1, which does not seem very important when compared to the paramount advantage, and therefore 10 IoA were assigned.

Step 7: Evaluate cost data

Table 6.3 presents the initial and life-cycle costs of these three alternatives. At the time, the PV cost calculations included available incentives from the local electric utility company, PG&E.

As we have shown before, the decision-making team can plot total IoAs against the initial cost (Figure 6.1), but they can also plot the total IoAs against life-cycle cost (Figure 6.2) for the three alternatives. This gives the team a better way to evaluate the decision by not only focusing on the short term. Figure 6.1 shows that the first cost follows almost a linear curve with IoAs. The design team needs to decide if they can and are willing to pay more to get alternative 3 with the highest IoA.

The more sustainable system was cheaper in the long term if one considers the museum's operating cost. Figure 6.2 shows that alternative 3 is better in the long term in life-cycle cost and IoA than alternatives 1 and 2. Since alternative 3 has the advantage of achieving the NZE goal and reducing CO_2 emissions to almost zero (for the operation of the HVAC system), it makes sense for the design team to select alternative 3, which was what the museum chose at the time. They were willing to invest in the HVAC system that helped them capture renewable energy (PV in the roof and bay water exchange) and

Table 6.3 Initial and life-cycle costs of the three alternatives ($)

	Alternative 1: *Natural gas boiler and chiller system with cooling tower*	*Alternative 2:* *Electric heat pump sourced by utility company with bay water*	*Alternative 3:* *Electric heat pump sourced by PV panel system with bay water*
Initial cost	1,154,250	1,654,250	6,173,399
Annual operation cost of energy	269,754	245,145	0
Annual maintenance costs	1,500	6,000	6,000
Replacement costs HVAC (every 40 years)	912,500	596,250	596,250
Life-cycle cost (50 years)	15,857,614	14,956,841	7,218,712

Source: Integral Group (2007), a group of green engineers.

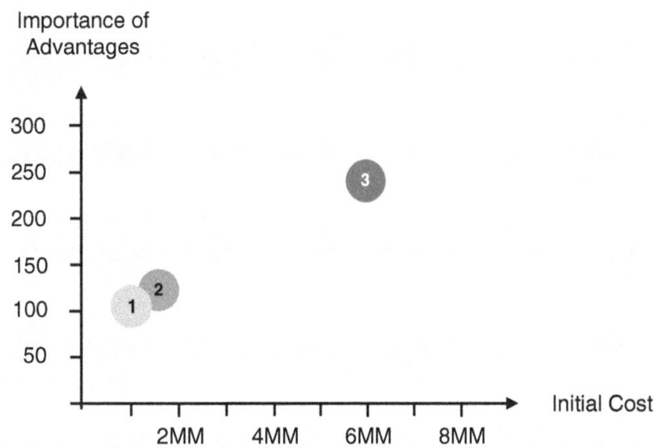

Figure 6.1 IoA vs. initial cost

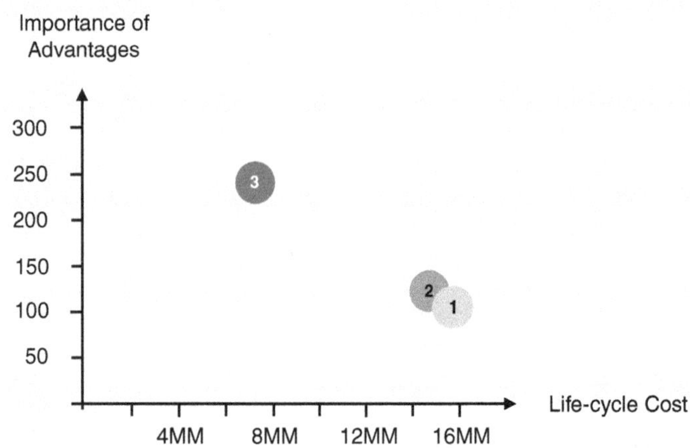

Figure 6.2 IoA vs. life-cycle cost

be more sustainable in the long term, even when that system needed more money upfront.

In this case, the museum was a nonprofit, and they raised more money to get the design they valued the most. It was easier to get donors because they were creating a net zero energy building, which appealed to donors in the San Francisco Bay Area, California. It was easier for donors to put money into something that their values aligned with. The solution they chose was more innovative and sustainable, too. Anecdotally, the museum was still satisfied with its decision to go to Net Zero Energy by 2024, and it became one of the first NZE museums in the world (Boyer, 2013).

Key points

- CBA can be used to organize the information transparently so all stakeholders understand why the decision is made. CBA helps the design team avoid making assumptions or basing the decision on previous experiences. Instead, the design team needs to understand the differences between the alternatives in the particular decision context, decide which alternative has advantages for each factor, understand how important these advantages are compared to each other, and finally, decide on the importance of the advantages based on a clear and shared rationale.
- CBA helps the design team build consensus by gathering relevant facts and then presenting trade-offs of cost vs. value. This case demonstrates how to incorporate multiple perspectives (e.g., owner, architects, mechanical engineer, operation manager, and users). Different perspectives can be discussed using factors, criteria, attributes, and an understanding of the advantages before estimating the IoAs. The major challenge in building consensus is when the design team has to agree on the IoAs.
- CBA helps document the rationale behind the decision-making process. This can help designers learn from past experiences and improve designs for future projects. In addition, documenting decisions with CBA helps retain knowledge of the project in case people leave the design team.
- This case shows how differentiating first from life-cycle cost is relevant to making long-term decisions, which is closely related to allowing for innovation and sustainability.

References

Arroyo, P., Tommelein, I. D., Ballard, G., & Rumsey, P. (2016). "Choosing by Advantages: A Case Study for Selecting an HVAC System for a Net Zero Energy Museum." *Energy and Buildings*, 111, 26–36.

Boyer, M. (2013). "San Francisco's Exploratorium Reopens as the Country's Biggest Net-Zero Energy Museum." Inhabitat. Avilable at: www.inhabitat.com/san-franciscos-exploratorium...

Lechner, N. (2008). *Heating, Cooling, Lighting: Sustainable Design Methods for Architects*. 3rd edn. Hoboken, NJ: John Wiley & Sons.

Suhr, J. (1999). *The Choosing By Advantages Decisionmaking System*. Westport, CT: Quorum.

Woolsey, K. (2012). "Sustain: The Museum as Exhibit." Exploratorium. Available at: www.exploratorium.edu

7 Not all decision-making methods are created equal

In this chapter, we will present the benefits of using Choosing By Advantages (CBA) from our research, comparing it against the most widely used methods in practice. If you do not understand other methods and the assumptions they make, you may think that CBA is just another method. However, here we can help you understand why CBA creates better conversations for decision-making, especially when we have to collaborate with multiple stakeholders who may have conflicting views. Not all methods are created equal! As we discussed in Chapter 4, we want a decision-making method: (1) to help create transparency in a way that teams can understand why the decision was made; (2) to support building consensus, even when there are different opinions, we want teams to build common ground and align on a direction; and (3) to allow teams to continuously learn if conditions change or similar circumstances occur in the future.

In this chapter we will introduce you to an example of how CBA differs from a widely used multi-criteria method called Weighting, Rating, and Calculating (WRC) that you might already be familiar with or are using without knowing its name or assumptions. Then, we discuss how both methods differ in achieving the three characteristics we want for a decision-making method (see Chapter 4). In addition, we present peer-reviewed published research to show the benefits of CBA over WRC. These studies used controlled experiments, inspired by Kahneman and Tversky's research on decision-making, where groups of decision-makers are given the same information, and asked to make decisions using the WRC method and then using the CBA method. One of the studies focuses on the experience of the decision-makers, and the other study focuses on the decision result. Both studies are fascinating to understand, and we hope they will give you the confidence to start your own CBA journey if you have not already started.

Another method that you may have seen applied in the construction industry is the Analytical Hierarchy Process (AHP); you may have been taught it in engineering school or perhaps read about it in a research publication. We will briefly introduce you to AHP and then present research that compares CBA to AHP using a case study to discuss its characteristics.

DOI: 10.1201/9781003518310-8

By the end of this chapter, you will understand what happens when you use WRC, AHP, and CBA methods. By showing you the flaws in both WRC and AHP assumptions, we hope to persuade you to stop using them and start using CBA.

Comparing CBA and WRC

Let's use a simple example to demonstrate the difference between CBA and WRC. Imagine you have to replace your household light bulbs, and you have to agree with your family or roommates on this decision. The decision is to choose one 800 lumens light bulb (60W equivalent) from the alternatives: (1) an incandescent light bulb (Figure 7.1), (2) a compact fluorescent (CFL) light bulb (Figure 7.2), or (3) a light emitting diode (LED) light bulb (Figure 7.3). For the sake of simplicity, this example assumes that each factor has only one criterion for evaluation and that decision-makers will choose the best alternative regardless of the cost.

Table 7.1 presents the factors and attributes of the alternatives. The attributes regarding the factor look are subjective, and we imagine they represent your opinion. The look of the light bulb is relevant depending on its use. For example, if the light bulb is installed in an enclosed lamp, the light bulb would not be visible, and therefore, the look is irrelevant. For this example, we assume the light bulb would be visible.

Figure 7.1 Incandescent light bulb
Source: Getleducated.com.

Figure 7.2 CFL light bulb
Source: Getleducated.com.

Figure 7.3 LED light bulb
Source: Inhabitat.com

Table 7.1 Attributes of light bulb alternatives

Factor	*Alternative 1: incandescent**	*Alternative 2: CFL*	*Alternative 3: LED***
Energy efficiency (Lumens[1]/ Watt)	14 lm/W	60 lm/W	64 lm/W
Readiness (Turns on instantly)	Turns on instantly	Turns on within a second and takes 30 to 60 seconds to achieve full brightness	Turns on instantly
Safety (mercury content)	No mercury	4 mg mercury/bulb	No mercury
Light quality (CRI[22])	100	82	93
Look	Nice	Very nice	Ugly

Notes: This example was developed in 2014 (Arroyo et al., 2014), and some of the alternatives we have now changed drastically:

* In some places, such as California or Europe, incandescent light bulbs are not available for purchase due to their low energy efficiency. This is a policy measure, given that people were still buying them just because they were cheap, which demonstrates the hyper-focus on the cost problem described in Chapter 3. However, we included them here to illustrate a decision.

** LED lighting has improved aesthetic choices over the years; now, we have LED light bulbs that look like incandescent bulbs. However, for the sake of this example, we are using only the LED light bulb shown in Figure 7.3.

[1] *Lumens:* The unit of measurement of the flow of light, or "luminous flux." With light bulbs it provides an estimate of the apparent amount of light the bulb will produce.

[2] *Color Rendering Index (CRI):* CRI represents the quality of light and its ability to render colors correctly. The maximum is 100, and represents the appearance of colors in daylight.

The Weighting, Rating and Calculating (WRC) method

The WRC method is colloquially known as the weighted sum model. The WRC method can be summarized in the following steps:

1. Identify the alternatives.
2. Identify the factors for evaluation.
3. Weight the factors.
4. Rate the alternatives for each factor (using an implicit criteria).
5. Calculate the value of each alternative and come to a final decision.

The WRC method example: choosing a light bulb

First, identify the alternatives (step 1), then identify the factors for evaluation (step 2). This example shows how the WRC method works when using a direct weighting of factors (step 3). In this case, decision-makers need to assign weights of factors, as shown in the second column of Table 7.2. The next step is rating attributes (step 4). Here, decision-makers create marginal

Table 7.2 Using the WRC method to choose a light bulb

Factor	*Weight %*	*Alternative 1*	*U*	*Alternative 2*	*U*	*Alternative 3*	*U*
		incandescent		*CFL*		*LED*	
Energy efficiency (Lm/W)	30	14 Lm/W	1.1	60 Lm/W	4.7	64 Lm/W	5.0
Readiness (Turns on instantly)	10	Turns on instantly	5.0	Turns on within a second, and takes 30 to 60 seconds to achieve full brightness	3.0	Turns on instantly	5.0
Safety (Mercury content)	30	No mercury	5.0	4 mg mercury / bulb	2.0	No mercury	5.0
Light quality (CRI)	20	100	5.0	82	4.1	93	4.7
Look	10	Nice	4.0	Very nice	5.0	Ugly	2.0
Total U	100		3.7		3.6		4.4

utility functions (column U in Table 7.2) to document how well an alternative performs on a given factor; for quantitative factors, such as energy efficiency and light quality, the numbers are calculated linearly using a linear scale to value the attributes from 0 to 5, where 5 is the best score (the rating applies an implicit criteria about what is better). The marginal utility functions for factors readiness, safety, and look were directly rated using a scale from 0 to 5. Finally, the total score is calculated using a weighted sum of factors and attributes scores (step 5). For example, the incandescent score will be:

30% of 1.1 + 10% of 5 +30% of 5+ 20% of 5+ 10% of 4 = 3.7

In this case, the highest score is alternative 3, the LED light bulb, with a score of 4.4.

Method assumptions and discussion

Some of the assumptions of the WRC method are:

- Marginal utility functions (UX1, UX2, ..., UXn) exist and usually are based on the assumption that attributes have a natural scale.

- Factors can be weighted. That means that decision-makers can prefer one factor (high-order abstraction concept) over another.
- Factors can be identified and weighted independently of alternatives' attributes.
- Cost can be a factor.

WRC is part of a family of multi-criteria decision-making (MCDM) methods known as value-based methods, which may have different impacts, depending on how the decision-making team constructs the weight of factors (w_i) and marginal utility functions. But does the WRC method help create transparency in the sense that everyone can understand why the decision was made? Does it help build consensus, meaning people a gree on what is important and can move forward with the decision result? Does it enable continuous learning by helping decision-makers make sense of new information as they learn more about the decision? As explained in Chapter 4, these are desired characteristics for a sound decision-making method. Let's discuss WRC characteristics in more depth.

- *Creating transparency*: The overall score of the alternatives provides a rationale for choosing an alternative. (1) However, when using fixed weights during the evaluation of alternatives, stakeholders assume that trade-offs between factors are fixed. For example, an increment in the marginal utility of light quality is always twice as valuable as an increment in readiness. However, this may not always reflect reality. There is no linear relationship between an increment on CRI and a reduction of time to turn on the light bulb. When using WRC, stakeholders make rules without really making sense of the value of increments. (2) In addition, assuming that every increment in the marginal utility of light quality, or any factor, is the same may not always reflect reality. For example, a difference in CRI from 80 to 90 may be more valuable than an increment from 90 to 100. (3) The differences between alternatives may not be highlighted if the factors that do not differentiate between alternatives are given high weights, which can mask the true difference between alternatives. (4) Cost can be a factor which allows for mixing cost and value of the alternatives. If that is the case, the design team may not be able to analyze value vs. cost. As discussed in Chapter 3, most likely, the cost will be given a high percentage weight, hiding differences even more and leading many people to choose the cheapest alternative.
- *Building consensus*: In this case, the decision-making team needs to reach a consensus on the weight of the factors and the ranking of attributes. (1) Agreeing on which factors have more weight may be a source of conflict, especially when team members have different interests. As discussed, factors do not represent specific judgments but general categories. (2) Factors can be weighted regardless of the attributes of the alternatives by basing the weights on previous experiences. Therefore, the team may miss the context of the particular decision. (3) In terms of managing

subjectivity in this case, the decision-making team would need to assign weights to factors first, which is a subjective task and it may contain biases; different people will argue for different weights of factors based on their own experience. Then, decision-makers need to rate the alternatives for each factor according to their attributes, which is also a subjective task. Although the calculation is objective, it is based on creating subjective linear scales.

- *Continuous learning*: This method provides an overall ranking of the alternatives. Therefore, (1) the decision-making team can have a score representing the value of the discarded alternatives. However, the advantages of the alternatives are not highlighted, and it can be difficult to remember later why we chose an alternative just based on a score; this is relevant if the conditions for the decision change and the decision has to be re-evaluated, as is the case in many projects. (2) If a new alternative is added, its attributes need to be rated, and the team may apply the same weights to factors. (3) When new factors are added, they need to be assigned a weight. This automatically results in a change of weights of the previous factors if the design team uses a fixed scale or rule (e.g., the sum of the weights of all factors has to be 100%). Then, a calculation is required to evaluate all the alternatives according to the new weights of the factors.

The rationale for choosing an LED light bulb using the example above can be documented as:

> The LED light bulb was chosen because it has a better overall score of 4.4 points vs. the CFL, which has 3.6, and the incandescent, which has 3.7 points.

That rationale is enough to support the decision. However, the differences between the attributes of the alternatives may not be highlighted.

The CBA method example: choosing a light bulb

Now, let's see how the same information will be used to make a decision with CBA; look at the CBA Tabular Method steps in Chapter 5, as based on Suhr (1999). Here, decision-makers describe advantages based on factors and criteria established in the first column of Table 7.3. Notice criteria are written explicitly on the table.

Instead of weighting the factors and rating the attributes of each alternative as in WRC, in CBA, decision-makers weight the advantages represented on a scale of importance of the advantage (IoA). In this case, using a scale of IoA from 0 to 100, where 100 is assigned to the most important advantage, also known as the paramount advantage (here 64 Lm/W vs. 14 Lm/W). In order to assign IoA to other advantages, decision-makers compare the advantages to

Table 7.3 Using the CBA method to choose a light bulb

Factor (Criterion)	*Alternative 1: Incandescent*		*Alternative 2: CFL*		*Alternative 3: LED*	
Energy efficiency (Lumens/watt)	Att.: 14 Lm/W		Att.: 60 Lm/W		Att.: 64 Lm/W	
(Higher is better.)	Adv.:	IoA:	Adv.: 46 LM/W higher than alternative 1	IoA: 90	**Adv.: 50 LM/W higher than alternative 1**	IoA: 100
Readiness	Att.: Turns on instantly		Att.: Turns on within a second, and takes 30 to 60 seconds to achieve full brightness		Att.: Turns on instantly	
(Turns on instantly is better.)	**Adv.: Turns on instantly vs. 30 to 60 seconds delay**	IoA: 10	Adv.:	IoA:	**Adv.: Turns on instantly vs. 30 to 60 seconds delay**	IoA: 10
Safety	Att.: No mercury		Att.: 4 mg mercury/bulb		Att.: No mercury	
(No mercury content is better.)	**Adv.: It has no mercury vs. 4mg/bulb**	IoA: 10	Adv.:	IoA:	**Adv.: It has no mercury vs. 4mg/bulb**	IoA: 10
Light quality	Att.: 100 CRI		Att.: 82 CRI		Att.: 93 CRI	
(Higher is better.)	**Adv.: 18 more CRI points than alternative 2**	IoA: 50	Adv.:	IoA:	Adv.: 9 more CRI points than alternative 2	IoA: 45
Look	Att.: Nice		Att.: Very nice		Att.: Ugly	
(Nicer is better.)	Adv.: Nicer than alternative 3	IoA: 10	**Adv.: Much nicer than alternative 3**	IoA: 20	Adv.:	IoA:
Total IoAs		**80**		**110**		**165**

Notes: Att. = attribute; Adv. = advantage. IoA = Importance of Advantages.
Underlining = least preferred attribute; bold = most important advantage in each factor

the paramount advantage. Finally, decision-makers can sum up the IoAs of each alternative and come to a conclusion. In this case, alternative 3 has the highest IoAs.

CBA method assumptions and discussion

CBA assumes that decision-makers can quantify their preferences. However, CBA requires decision-makers to identify the advantages of alternatives prior to constructing their preferences:

- CBA assumes that advantages can be weighted based on pairwise comparisons with other advantages. Therefore, it is possible to aggregate advantages.
- In contrast with most other methods, CBA does not require factor weights (wi) or attribute weights (U). Trade-offs are made directly among advantages.
- Cost cannot be a factor.

The assumptions of this method have several consequences:

- *Creating transparency*: (1) Decision-makers can provide a rationale for the decision and state what the trade-offs were among attributes of the alternatives. (2) The differences between alternatives are used to construct the advantages, and those are highlighted. (3) Cost cannot be a factor, and it is possible to make an analysis of value vs. cost.
- *Building consensus*: In this case, the decision-making team builds consensus little by little, by first agreeing on criteria per each factor, then agreeing on attributes, describing the advantages, and finally reaching a consensus on the weighting of advantages. In this case, specialists do not have to fight over which discipline (often represented by a factor) will get a higher score; in this case, experts can work to provide relevant facts documented as attributes. (1) This method helps in building consensus when the team agrees on the advantages of the alternatives, based on the difference between their attributes. However, it may be challenging for some teams to agree on the importance of the advantages, which requires having collaborative conversations. (2) The decision-making team is more likely to base decisions on the attributes of the alternatives, which are relevant facts based on the context of the decision. (3) In terms of managing subjectivity in CBA, the design team compares known attributes first, which is an objective task, and then weights the advantages, which is a subjective task.
- *Continuous learning*: (1) This method provides an overall ranking of the alternatives and provides advantages of all the alternatives; if conditions change or new alternatives are created, the team can take ideas from the advantages of the discarded alternatives, and create a better hybrid alternative. (2) If a new alternative is added, the design team needs to describe

> and assign a weight to its advantages. (3) If a new factor and criterion are added, the decision-making team needs to assess which alternatives have advantages for that factor, and weight the importance of these advantages; no recalculations are needed, just assessing the importance of the new advantages against the paramount advantage. (4) Multiple decisions can be compared using this method if the scale of importance is aligned. This is particularly useful in a project where many design alternatives will be evaluated for different areas (e.g., renovating the kitchen, building a garage, and installing solar panels in a house) to help decide what is the best way of spending a budget.

The rationale for choosing a LED light bulb using the example above can be documented as:

> The LED light bulb provides more important advantages than the CFL and incandescent lights. The LED light bulb has 50 lm/W more than the incandescent one (65 lm/W vs. 14 Lm/W); the LED light bulb turns on instantly vs. the CFL turns on within a second and takes 30 to 60 seconds to achieve full brightness; the LED light bulb has no mercury vs. the CFL has a 4 mg/bulb; and the LED light bulb has 93 CRI vs. the CFL has 82 CRI.

The advantages of the discarded alternatives are also highlighted:

> The CFL bulb has 46 lm/W higher than the incandescent one (46 lm/W vs. 14 Lm/W); and the CFL bulb is much nicer than the LED bulb. Finally, the incandescent light bulb turns on instantly vs. the CFL turns on within a second and takes 30 to 60 seconds to achieve full brightness; the incandescent light bulb has no mercury vs. the CFL has 4 mg/bulb; the incandescent light has 100 CRI vs. the CFL has 82 CRI; and the incandescent light bulb is nicer than the LED bulb.

In this way, decision-makers can highlight all the relevant differences. In CBA the attributes of the alternatives are not translated to any other scale. The alternatives are compared using the original attributes units such as CRI or mg/bulb.

Comparing CBA and WRC: focus on decision-makers' experience while collaborating

Now that you understand the methodological difference between WRC and CBA let's see if the methods differ and how they impact teams' decision-making. Arroyo et al. (2016) compared CBA and WRC methods in terms of how they support group collaboration, particularly concerning design decisions involving sustainability factors in the construction industry. Before this study, the impact of using different methods had not been tested using a controlled experiment. This is particularly true in the context of building design decisions. We only had case studies comparing WRC and CBA (Parrish and Tommelein, 2009; Arroyo et al., 2012; 2014), such as the example of the lightbulb explained in the previous section.

This study is based on an experiment Paz led while she was a professor in Chile, where researchers invited 15 architecture, engineering and construction (AEC) practitioners to take part. They were architects and engineers between 30 and 45 years old and had five or more years of professional experience in the construction industry. The participants were arranged into five groups of three people each. They had to play a role in the decisions as owners, architects, or engineers. The researchers tried to assign their roles according to their professions, although it was not possible in all cases. All groups made the same 12 decisions pertaining to choosing alternative materials, components, building systems, building layouts, and building locations. The experiment used multiple baseline testing (MBT), a popular research method used to test the impact of an intervention, sometimes used for testing drugs in the pharmaceutical field. In this experiment, all groups made 12 decisions (Figure 7.4). Practitioners made the initial decisions using the WRC method, which everyone was familiar with from previous experiences, such as calculating a weighted average, a standard mathematical exercise students learn at school. Later, in a staggered manner between the fifth and ninth decision, the groups learned to apply the CBA method. In this study, the CBA method is considered to be the intervention in the experimental design.

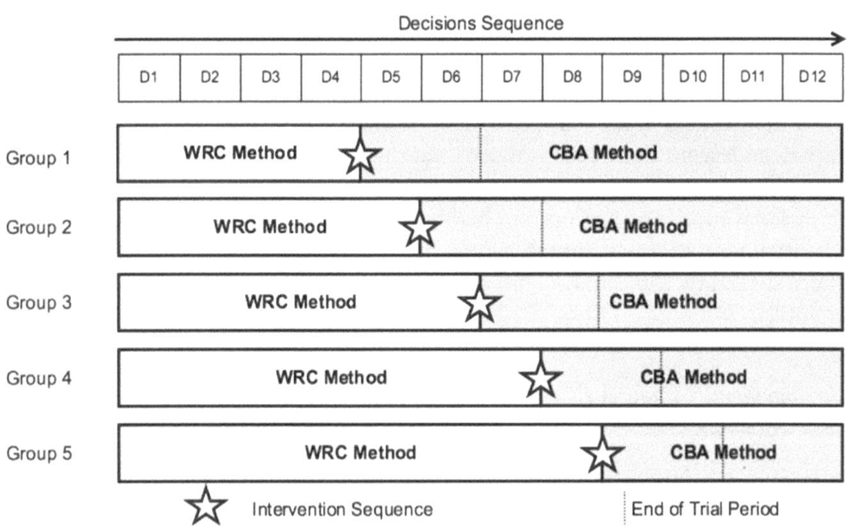

Figure 7.4 Experimental design for MBT to compare WRC and CBA
Source: Arroyo et al. (2016).

Four variables were analyzed: (1) time to reach consensus (researchers timed each group), (2) satisfaction with the final decision, (3) personal frustration during the decision-making, and (4) perceived frustration from others during the decision. To know about satisfaction and frustration, we asked each participant to fill out a survey (measured on a scale from 0 to 20, where a higher score represented more satisfaction or frustration) after each group decision was made; we also asked at the end of the experiment which method they would use in future decisions.

The results showed the following results (Arroyo et al., 2016):

1. CBA was faster with statistical significance than WRC at reaching consensus. Across all groups, the groups took 13.02 minutes on average to reach a consensus when WRC was adopted. However, this dramatically reduced to 6.32 minutes when the CBA method was adopted, signifying a time savings of over 6 minutes for each decision.
2. The results do not show statistical support in favor of any method regarding satisfaction with the final decision. According to the findings, practitioners were not used to making decisions using any structured method. Even though they were familiar with WRC conceptually, they were not using it to make collaborative decisions on their projects. Therefore, they valued the support of both methods in terms of satisfaction with the final decision.
3. Participants reported less personal frustration using CBA than WRC with statistical significance. Comparing across all groups, the frustration level when WRC was adopted was 5.49. The overall frustration level was reduced by 2.36–3.13 after researchers introduced CBA as the intervention.
4. Participants reported less perceived team frustration using CBA than WRC with statistical significance. Across all groups, the perceived frustration level was 5.74 when using WRC and 3.48 when using CBA.

Finally, 86% of all participants chose CBA rather than WRC for future application in decision-making. The main reason given by the respondents was that CBA makes decisions more objectively and reduces the conflict among the participants. They highlighted that CBA provides more transparency in identifying the differences between the alternatives, allowing the decision-makers to base their decision on those differences and not on subjective assessments as WRC does. These results were relevant because they show that CBA supports collaboration better than traditionally used methods, such as WRC.

CBA vs. WRC: focus on decision results in hotel design research

Testing whether or not a decision result is better than others is not simple. As stated in the previous study, practitioners did not show statistical support for any method regarding satisfaction with the final decision. However, that does not mean that there is no difference in the results if different decision-making methods are used. The idea and the opportunity to test CBA and WRC results came a few years later.

Around 2015, while Paz was a full-time professor in Chile, she learned that the Center for Integrated Facility Engineering (CIFE) researchers at Stanford University were implementing a relatively new technology called Parametric Design, which allows you to set the parameters of a building and create multiple designs. Researchers can explore alternatives for buildings and consider all the different combinations of alternatives you could have for your building. CIFE researchers were changing the ratio of windows to walls, the orientation of the building, and different types of materials so they could test thousands of alternatives and ideally expect to have a better design than using traditional design methods; this is still a very exciting technology with great potential. In a presentation one of the CIFE researchers was giving in Chile, she asked the presenter how they made the trade-offs, and how they knew what the best alternative was. At the time, there was no clear answer, so they started a research project together that led to very interesting findings and resulted in a conference and journal publication a few years later (Correa et al., 2017; Arroyo et al., 2018).

For this new study, the researchers designed an experiment and asked a group of practitioners to use WRC and CBA to choose the optimal solutions; in this case, they measured whether or not practitioners obtained the best value for the budget they chose to spend in the design of a hypothetical hotel in Orlando, Florida. Using parametric design, the researchers used several parameters to create 1,000 design alternatives and then identified factors to evaluate the alternatives.

Practitioners first used WRC by weighting factors, and researchers applied those preferences and rated the alternatives using the attributes obtained by the model and calculated a result. Then the practitioners used the CBA Tabular method to choose a design. Obviously, they did not make a table with 1,000 alternatives; that would make CBA impractical, and the study would probably get no research volunteers. So the researchers had to innovate and create a new method to evaluate a large number of design alternatives and keep CBA principles true (Arroyo et al., 2018). Basically, researchers chose extreme alternatives and then asked practitioners to make a decision using the CBA Tabular method for a reduced number of alternatives (between three to five), and then researchers extrapolated the preference curves. In this way, they ensured the decision was made based on the importance of advantages.

What the researchers found was that when using WRC, most construction practitioners, as expected, weighted costs very high compared to other factors that represented value (remember hyper-focus on cost discussed in Chapter 3), such as VOC emissions, energy usage, and acoustic performance. As a result, by using WRC, practitioners made sure that they never chose any of the most expensive alternatives. However, the most surprising result is that they often chose alternatives that were not the best they could get for the cost they settled for. Researchers could demonstrate this afterwards when practitioners used CBA with the same information to develop a graph of cost versus IoA on a separate axis.

Figure 7.5 Sub-optimal decision example
Source: Correa et al. (2017).

Figure 7.5 shows the result of a practitioner's decision; each dot represents a design alternative, showing the evaluation of value, measured as the IoA (y-axis) and cost of building the design (x-axis), the large bold dot in the center is the chosen alternative with WRC. As you can see for that same cost several alternatives provide more value (if you move in a straight line up); also, many alternatives are less expensive and provide the same value (if you move in a horizontal line to the left); moreover, alternatives in the right upper quadrant from the chosen alternative will be better (weather, less expensive, more valuable, or both). As you can see, this is a sub-optimization and a real example of what happens when decisions are hyper-focused on cost, and there is no way to differentiate cost from value. This sub-optimization may happen to many more decisions when we use the popular WRC method. On the contrary, the research showed that when practitioners were taught and used the CBA method, they would always choose the best value for the cost they settled on. CBA makes this a simple decision by seeing what they get for the cost they are willing to spend. They chose optimal alternatives. Figure 7.6 shows an example where a team chose an optimal alternative (in bold); no other less expensive alternatives provide the same or more value.

In summary, when using WRC, the experiment participants hyper-focused on cost. In many cases, they did not get the most value for the price they were willing to pay (or invest in the design of the building). They were making

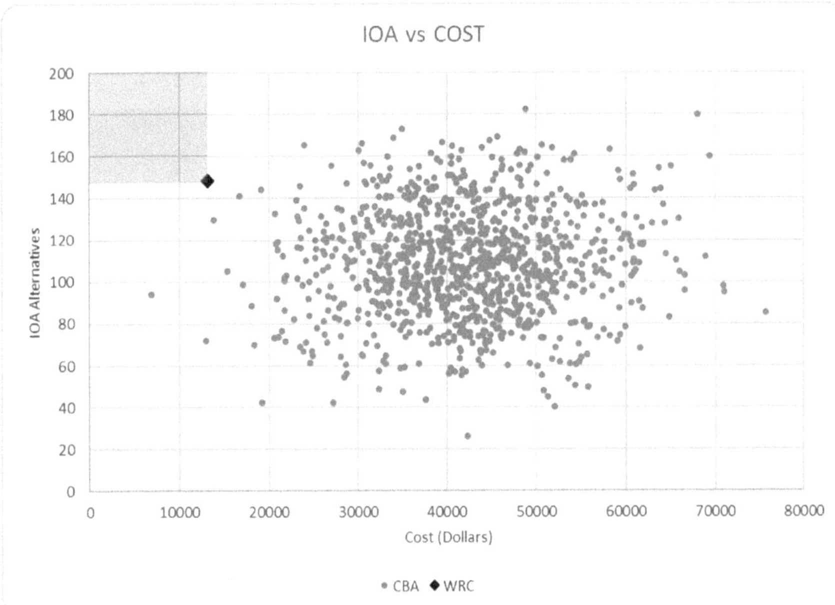

Figure 7.6 Optimal decision example
Source: Correa et al. (2017).

assumptions, such as using linear preferences and creating rules that they probably did not fully understand, such as weighting cost very high, thinking that this would ensure they would limit the money they were willing to spend. This strategy only ensures that the most expensive alternatives do not get a good score. However, they did not have the vision to see value versus cost, so they could not choose the best alternative for the price they were willing to pay.

Comparing CBA and AHP

Now that you understand WRC, we will introduce the AHP, which is a more complicated decision-making method than WRC, but it shares some of the same assumptions. The procedure for using the AHP can be summarized as follows (Saaty, 2008):

1. Model the problem as a hierarchy containing the decision goal, the alternatives for reaching it, and the criteria for evaluating the alternatives.
2. Establish priorities among the factors by making a series of judgments based on pairwise comparisons of the elements.
3. Establish priorities among the alternatives for each factor based on pairwise comparisons of attributes.

4. Synthesize these judgments to yield a set of overall priorities for the hierarchy.
5. Check the consistency of the judgments.
6. Come to a final decision based on the results of this process.

Step (2) measures the relative importance of factors and alternative preferences through pairwise comparison matrices, which are recombined into an overall rating of alternatives using the eigenvalue method. When weighting factors, decision-makers are asked to indicate the strength of their preferences for one factor over another on the following scale: 1 equally preferred, 3 weak preference, 5 strong preference, 7 demonstrated preference, and 9 absolute preference.

Example: choosing an exterior wall assembly

An example compares different exterior wall assemblies for a residential building in California. For simplicity, we are comparing two alternatives: (1) standard wall (Figure 7.7), and (2) double stud wall construction (Figure 7.8).

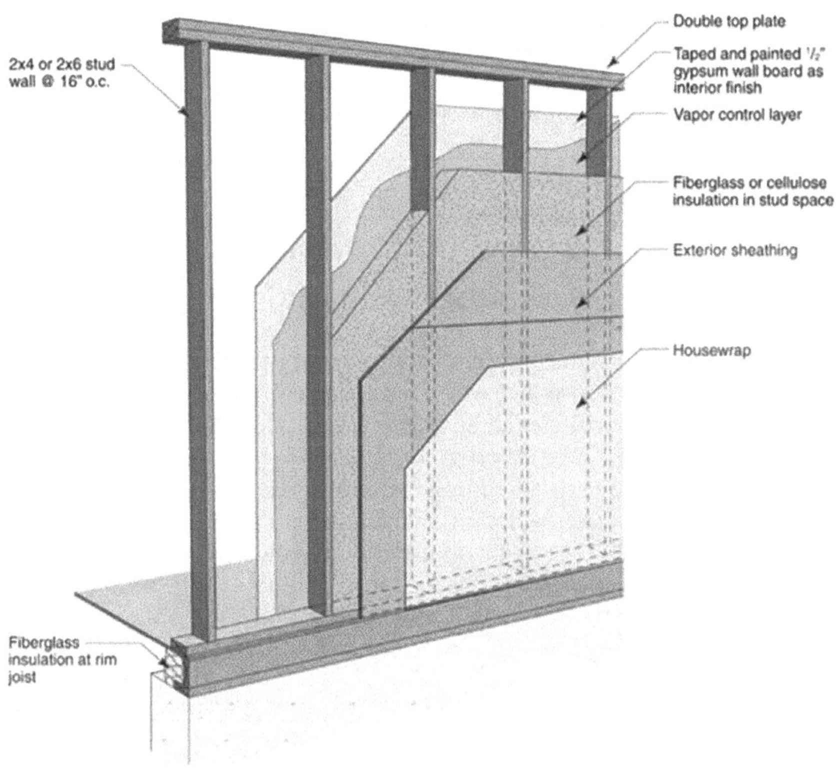

Figure 7.7 Standard wall design
Source: Building Science Corporation (2009).

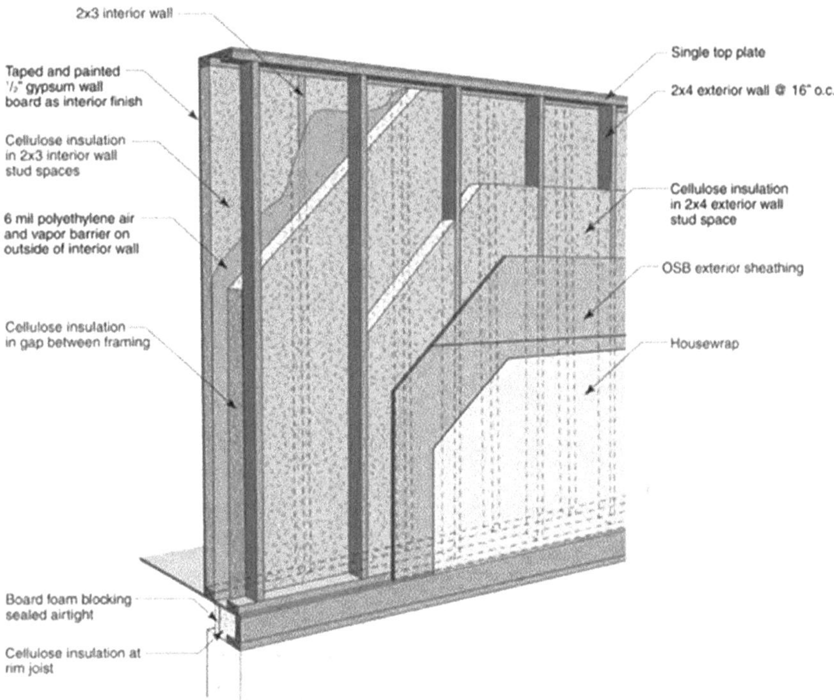

Figure 7.8 Double-stud wall design
Source: Building Science Corporation (2009).

In this example, we considered the following factors and criteria:

1. *Thermal control* relates to the amount of energy that the building needs to maintain in order to achieve a thermal comfort level for its occupants. This is expressed in a characteristic measured through the R-value of the whole-wall system. The criterion for selection is "the higher R-value, the better," as this means higher insulation properties.
2. *Durability* relates to how long the building will last and depends on the ability of the wall to stop rain, moisture, and air leakage. The criterion for selection is "the building must last 50 years."
3. *Buildability* refers to how easy it is to build the wall assembly considering the current available knowledge. The criterion for selection is "the easier to build, the better."
4. *Material use* refers to the quantity of material used for wall assembly, and it is related to the embodied energy of the wall, which certainly also depends on the type of material used for insulation. The criterion for selection is "the less material used, the better."

Table 7.4 Step 2 of AHP

Factors	*Thermal control*	*Durability*	*Buildability*	*Material use*
Thermal control	1	3	1/3	1/5
Durability	1/3	1	3	3
Buildability	3	1/3	1	1/5
Material use	5	1/3	5	1

AHP: choosing an exterior wall assembly

Here, we present step (2) factor weighting of the AHP method. Table 7.4 presents the matrix of value judgment for the factors. The meanings of these numbers are the value preferences of the factors expressed as a ratio. For example, number 3 in the second column and first row of the matrix means that thermal control has a weak preference over durability, and correspondingly the first column's second row has a value of 1/3.

This process requires a high level of abstraction. The questions here are: What is more important, buildability or material use? Thermal comfort or durability? And so on. It is hard to defend that thermal control is more important than durability without considering what the relevant differences are between the alternatives. Finally, when the eigenvalue of this matrix is calculated, it provides the factor's weight, assuming linear trade-offs between the performances of the alternatives. As we have discussed before, that assumption is not always true; for example, stakeholders may not want the building to last more than 50 years, and if an alternative presents durability higher than that, stakeholders may not be willing to have a worse performance in buildability.

AHP method assumptions and discussion

Several of the assumptions of value-based methods apply to AHP:

- Marginal utility functions (UX1, UX2, ..., UXn) exist and usually are based on the assumption that attributes have a natural scale.
- Factors can be weighted. That means that decision-makers can prefer one factor (high-order abstraction concept) over another.
- Factors can be identified and weighted independently of alternatives' attributes.
- Cost can be a factor.

Assumptions specific to AHP are:

- Quantitative attributes can be normalized, assuming a natural ratio scale. We will not show the calculations in this example, but in step 3 of the AHP

process, you have to normalize the attributes of each alternative across each factor. This means forcing a scale from 0 to 1, and that the sum of the attribute scores of the alternatives is 1. For example, the standard wall's buildability score can be 0.7, and the double-stud wall's score can be 0.3.
- Factors are independent, so the conversion rates (wi) between factors (when integrating the utility function) are constant.

The problem with both WRC and AHP methods is that the process of weighting factors is subjective. When using these methods, stakeholders need to decide what is more important: economic, social, or environmental aspects of sustainability. This leads to conflicting questions, such as: what is more important when choosing a material: impacts on human health, embodied energy, or life-cycle cost? This question has a high level of abstraction; it does not have a particular meaning for a given decision context. By using this approach, the differences between the alternatives are not highlighted, and therefore, decisions are not anchored to relevant facts. The method may not help in creating transparency, building consensus, and continuous learning if the team directly assigns weights to factors or if they assume linear marginal utility functions. Some of the issues include:

- *Creating transparency*: (1) In AHP, the weight of the factors is calculated using an eigenvalue vector, which makes it hard for most people to understand and, thus, trust in the decision outcome. In addition, the result provides fixed weights of the factors, which means that stakeholders assume that trade-offs between factors are fixed. For example, an increment in the marginal utility of thermal control is always twice as valuable as an increment in buildability. However, this may not always reflect reality. (2) In addition, assuming that every increment in the marginal utility has the same value may not always reflect reality. AHP also forces attributes to be normalized, so scores are from 0 to 1. (3) The differences between alternatives are not highlighted if factors that do not differentiate between alternatives are given high weights, which can mask the true difference between alternatives. (4) Cost can be a factor that allows for the mixing of cost and value of the alternatives. If that is the case, the design team may not be able to make an analysis of value vs. cost. Moreover, Arroyo et al. (2015) carried out an experiment that highlighted the intrinsic flaws of AHP by showing that by removing irrelevant information, a factor that did not differentiate the alternatives (both alternatives had the same attribute in that factor), the result of a decision changed. Yes, this is how blindly the algorithm works. Basically, AHP forces linear trade-offs and the creation of linear scales for everything, creating a complicated way of weighting factors, which a lot of engineers like because it gives the illusion of being objective. However, it's not objective; you are only making linear assumptions to make trade-offs.
- *Building consensus*: In this case, the decision-making team needs to reach a consensus on the preferences of each factor by making pairwise

comparisons, and the ranking of attributes is done by assuming a linear natural scale, from 0 to 1, regardless of the units of the attributes. (1) Agreeing on which factors are more important may be a source of conflict, especially when members of the team have different interests. As discussed, factors do not represent specific judgments but general categories. (2) Factors can be weighted regardless of the attributes of the alternatives by basing the weights on previous experience. Therefore, the team may miss the context of the particular decision. (3) In terms of managing subjectivity in this case, the decision-making team would make several small comparisons among factors. However, all of those are subjective tasks. Then, they would need to calculate a value from 0 to 1 by normalizing the attributes of each alternative, which is also a subjective task. However, this makes assumptions about linear preferences. Although the calculation is objective, it is based on subjective scales, and agreeing on subjective statements is an easy way of adding unnecessary conflict to a decision.

- *Continuous learning*: This method provides an overall ranking of the alternatives. (1) The decision-making team can have a score representing the value of the discarded alternatives. However, the advantages of the discarded alternatives may not be highlighted. (2) If a new alternative is added, its attributes need to be rated and the normalization of attributes of all alternatives has to be recalculated. (3) When new factors are added or deleted, they need to recalculate the weights, and inconsistencies can occur. As stated before, a decision can change even when you remove factors that are irrelevant to the decision (Arroyo et al., 2015), which makes continuous learning almost impossible with this method.

Table 7.5 Summarize the attributes of each alternative (step 4 in CBA Tabular method)

Factors (Criteria)	*Attributes of standard wall construction*	*Attributes of double stud wall Construction*
Thermal Control (The higher R value the better.)	R-10 (2x4 wall with R-13 stud space insulation)	R15 (2x4 wall with fiberglass batt)
Durability (It must last 50 years.)	Depend on exterior barrier.	Depend on exterior barrier.
Buildability (The easy the better.)	Easy to construct. Designers, trades and SC are use to it.	Not very complicated, but It requires custom frames for penetrations.
Material use (The less the better.)	Framing lumber could be minimized further if advanced framing was used.	Wall framing material is increased significantly due to secondary interior wall.

CBA: choosing an exterior wall assembly

Here, we are not going to introduce CBA again or repeat the assumptions behind it, but we will present the attributes of each alternative, step (4) and the advantages of each alternative, step (5) of the CBA Tabular method (see Chapter 5). Deciding the importance of each advantage, step (6) and evaluating cost data, step (7), are discussed but not developed because they require a subjective value judgment that will depend on the stakeholders involved in the decision-making process and the context of a particular building. A summary of the attributes of each alternative (step 4) is presented in Table 7.5.

According to the CBA Tabular method, the advantages of each alternative must be highlighted (step 2), which are considered the relevant facts for decision-making. Table 7.6 presents the advantages of each alternative relative to the least-preferred one.

Now, steps (4) and (5) of the CBA are objective and easy to agree with. However, stakeholders need to agree on criteria, which will depend on climate conditions, building orientation, building type, who uses the building, etc. Once the advantages are decided, CBA leads to subjective questions anchored to the relevant facts, such as: what is more important: the advantage in buildability and material use of the standard wall construction or the advantage in thermal control of the double-stud wall construction? It is important to note that the durability factor is irrelevant in the decision because no alternative has an advantage over another. Again, in WRC and AHP, this irrelevant information would have been assessed and impacted the decision outcome. It is essential to note that money (whether it is cost or price) is neither a factor nor a criterion. Money is treated separately in CBA as a constraint. Consider for this example the estimated first cost for the standard wall construction to be about $50/m^2 and for the double-stud wall construction to be about $80/m^2. When considering cost, the relevant question is: if stakeholders value the

Table 7.6 Decide the advantages of each alternative (step 5 in CBA Tabular method)

Factors (Criteria)	*Attributes of standard wall construction*	*Attributes of double stud wall Construction*
Thermal Control (The higher R value the better.)		R value is higher by 5 than standard wall
Durability (It must last 50 years.)		
Buildability (The easy the better.)	Easier to construct than double stud	
Material use (The less the better.)	It uses less material than double stud	

advantage of thermal control more, do stakeholders want to pay $30 more per m^2? Do stakeholders have the financial power to pay for that higher first cost in the short term? CBA allows for the flexibility to add more factors or alternatives without impacting the previous assessment of alternatives, in contrast to value-based methods, in which the score and weighting of factors must be recalculated any time a new factor is added to the decision. Some factors that can be added to the analysis are embodied energy of materials, aesthetics, etc. In addition, other alternatives may be structured insulated panel systems, truss walls, etc.

Key points

- CBA methods are superior to value-based methods (WRC and AHP) to support collaboration by driving transparency, consensus, and continuous improvement.
- CBA helps stakeholders make decisions based on relevant facts, focusing on the importance of the advantages between the attributes of the alternatives, which minimizes unnecessary conflicts. In contrast, value-based methods make stakeholders weight factors; regardless of any facts, they have to agree on what is more important in the abstract. This leads to more conflict, as people tend to defend what they represent (their department or function) or argue, based on previous experiences that may or may not be relevant to the decision context. Therefore, stakeholders may have difficulties resolving conflicting interests and collaborating.
- Methods that weight factors do not make decisions based on the relevant facts. They are also more likely to result in suboptimal decision outcomes, as they typically mix cost with assessments of value, as demonstrated in the hotel design experiment. Unfortunately, some sort of WRC is mandatory for many public governments to make decisions in public procurement (see Chapter 8); therefore, we live with the unintended consequences of focusing too much on cost. *You don't get the best value because you never understood value vs. cost!* Unfortunately, this is not just a construction problem. We know that in most organizations, when decision-makers assess value, they also mix it with cost, so we sub-optimize our resources and make poor decisions.
- We recommend using the CBA method instead of WRC, AHP, or any other method that requires weighting factors or attributes that do not focus on relevant facts. As we review in the research discussed in this chapter, *the CBA method will help practitioners make decisions in less time and with less frustration than WRC*. It will make the decision-making process more transparent and collaborative. This is especially relevant in the sustainable design of construction projects, where the design team makes hundreds of decisions, considering social, environmental, technical, and economic factors. Using the wrong method has consequences for the construction

industry. In Chapter 9, we explore the topic of how decisions impact sustainability.

- CBA allows for creating value for stakeholders while reducing decision-making uncertainty. This will reduce the amount of waste generated by unsound decisions, which aligns with lean thinking, including early collaboration. In Chapter 10, we will discuss how to set up the decision-making process for collaboration and make it efficient.

References

Arroyo, P., Fuenzalida, C., Albert, A., and Hallowell, M.R. (2016). "Collaborating in Decision Making of Sustainable Building Design: An Experimental Study Comparing CBA and WRC Methods." *Energy and Buildings*, 128, 132–142.

Arroyo, P., Mourgues, C., Flager, F., and Correa, M.G. (2018). "A New Method for Applying Choosing By Advantages (CBA) Multicriteria Decision to a Large Number of Design Alternatives." *Energy and Buildings*, 167, 30–37.

Arroyo, P., Tommelein, I.D., and Ballard, G. (2012). "Deciding a Sustainable Alternative by 'Choosing By Advantages' in the AEC Industry." In *Proceedings of the 20th Conference of the International Group for Lean Construction (IGLC)*, San Diego, CA, pp. 41–50.

Arroyo, P., Tommelein, I.D., and Ballard, G. (2014). "Comparing Weighting, Rating and Calculating vs. Choosing By Advantages to Make Design Choices." In *Proceedings of the 22nd Annual Conference of the International Group for Lean Construction*, Oslo, Norway, Vol. 1, No. 510, pp. 1–12.

Arroyo, P., Tommelein, I.D., and Ballard, G. (2015). "Comparing AHP and CBA as Decision Methods to Resolve the Choosing Problem in Detailed Design." *Journal of Construction Engineering and Management*, 141(1), 04014063.

Building Science Corporation. (2009). "High R-Value Wall Assemblies" Available at: http://www.buildingscience.com/documents/information-sheets/high-r-value-wallassemblies. (accessed April 8, 2012).

Correa, M.G., Arroyo, P., Mourgues, C., and Flager, F. (2017). "Comparing Choosing By Advantages and Weighting, Rating and Calculating Results in Large Design Spaces." In *Proceedings of the 25th Annual Conference of the International Group for Lean Construction*, pp. 259–266. doi.org/10.24928/2017/0248

Parrish, K., and Tommelein, I.D. (2009). "Making Design Decisions Using Choosing By Advantages." In *Proceedings of the 17th Annual Conference of the International Group for Lean Construction*, Taipei, Taiwan, pp. 501–510.

Saaty, T.L. (2008). *Decision Making for Leaders: The Analytic Hierarchy Process for Decisions in a Complex World.* Pittsburgh, PA: RWS Publications.

Suhr, J. (1999). *The Choosing By Advantages Decisionmaking System.* Westport, CT: Quorum.

8 CBA and the tendering procedure of construction projects

The media constantly reports on failed projects, cost overruns, and massive delays. Why is this happening? Are people in the construction sector incompetent, or is it a system-based problem? What are the underlying causes for these failures? Well, the result of a project depends on how it will be delivered. We have different organizations that need to work together to make a project work, and if the system does not support the participants' collaboration, the project might end up as a disaster.

Thus, one of the most essential decisions in architecture, engineering and construction (AEC) projects is how to deliver a project. This also includes the decision of how to integrate companies to design and build the project, represented in the tendering procedure. Public projects are primarily delivered traditionally, meaning they are bid out based on price or with a minor acknowledgment of factors such as quality. Tendering is the first interaction between an owner and the potential partner, so it needs to be considered that it sets a tone for how the parties will work together (Schöttle, 2022). It is not news that competitive bidding based solely on price too often results in changes, disputes, claims, and an unhealthy project organization. Many authors have stated this before (e.g., Elfving et al., 2005; Eriksson and Laan, 2007; Schöttle and Gehbauer, 2013; Rooke et al., 2004; Rosenfeld, 2014), so why do we only focus on price and how can CBA help to create a better tendering procedure?

Project delivery systems

There are several ways to deliver a project based on how the different parties are contracted, how the risk is placed, and how the teams are selected. The leading delivery systems are Design-Bid-Build (DBB), Construction Management at Risk (CM@Risk), Design-Build (DB), Integrated Project Delivery (IPD), and Project Alliancing (PA) (see Figure 8.1).

Public projects are primarily delivered using DBB. This includes a tendering procedure based on price or with a minor acknowledgment of factors such as quality. Here, each contractor is aligned with a single and traditional contract either to the owner or to another contractor (Chih, 2010, p. 13; Carpenter and

DOI: 10.1201/9781003518310-9

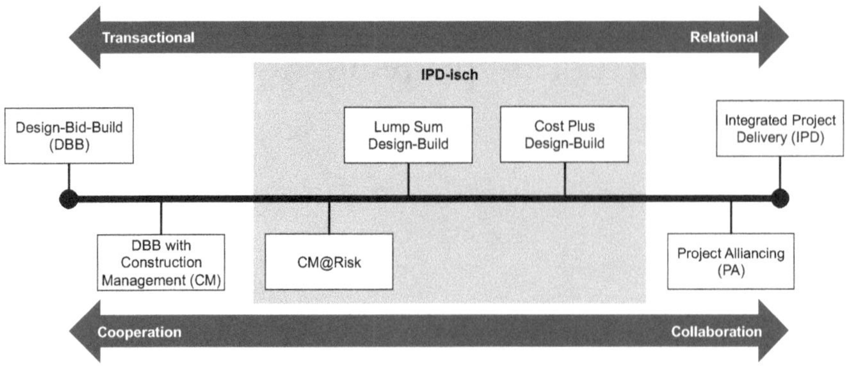

Figure 8.1 Overview of the leading project delivery systems
Source: Schöttle (2022).

Bausman, 2016), shifting risk from one party to another. Parties are selected in sequence and shortly before their work packages start. For example, the owner first selects the architect and the design engineers so they can start with the development of the design. Shortly before breaking ground, the owner tenders the groundbreaking company, followed by the general contractor and sometimes significant trades. The general contractor then sequences out the trades. As a result, design and execution are separated from each other, and there is a lack of an overlapping view (Tsao et al., 2000). The project is then managed sequentially with a reactive control mechanism (Ballard, 2012) by checking the work status. "[T]raditional contracts are designed to maintain distance between project members and insulate them from each other. Profitability is determined individually and is unrelated to project outcome" (Fischer et al., 2017, p. 363), leading to information asymmetry between the different organizations. Ultimately, every involved organization is looking out for its own interest, which is counterproductive when generating value for the customer/owner.

Where in DBB, a general contractor can be contracted to coordinate and manage the execution phase. However, in CM@Risk, a construction manager will coordinate the design phase, but contracts are still kept separated from each other as well as the design and execution phase (Chih, 2010; Forbes and Ahmed, 2010). In contrast, in DB, one party is responsible for the design and execution phase (Konchar and Sanvido, 1998; Wardani et al., 2006; Eriksson and Westerberg, 2011). The DB team usually consists of an architect, the general contractor, and significant design engineers and trade contractors.

There are ways of delivering a project with the approach of collaboration and shared risk, like IPD and PA. In both delivery systems, a multi-party contract among the key stakeholders is installed, including financial incentives. IPD emerged in the US private building sector (Johnson et al., 2013, p. 481), as people understood that there is no transparency if the interests of the different organizations are not aligned. In comparison, PA was developed in the Australian public sector as they faced the issue of managing major infrastructure projects (Love et al., 2011; Lahdenperä, 2012). Due to their origins, IPD and PA differ in their intention and, therefore, in the setting-up of the project (Schöttle, 2022). While PA creates an organizational framework to minimize risks potentially arising from the process itself, IPD intends to involve the knowledge of experts in the project as early as possible (Lahdenperä, 2012). The different intentions led to a different focus. For example, PA focuses on the tendering procedure. It defines precisely how companies participate (Heidemann and Gehbauer, 2010) but without questioning the methods used to evaluate proposals; in comparison, IPD focuses on production planning and creating a network of commitments, and contracts are developed jointly (Heidemann, 2011). So, why do public owners primarily use DBB if other more collaborative delivery methods exist, such as IPD and PA? It is challenging for a public owner to develop multi-party contracts based on the current legislation in most parts of the world, but that doesn't mean this cannot change. Nowadays, both delivery systems, IPD and PA, are merging together as people learn from practical experiences implementing them. However, when differentiating based on the approach, it can be stated that IPD contains a higher degree of collaboration than PA (Schöttle, 2022).

Owners start with the decision of which kind of delivery system they want to use to deliver the project. *CBA can help to better understand the project needs as well as the needs of the owner*. So, let's first take a deeper look into a decision from a project where the owner first decides if the project should be delivered using DBB or IPD and, second, decides which work scope should be part of IPD. Using CBA to answer the decision question also had the aim of getting a shared understanding and being able to develop a transparent and solid argument for the organization board to change the delivery system for the project.

CBA can help to create a shared understanding and a strong argument

In Munich, a core line of the city train connects the busiest part of the public infrastructure system. Opened in 1972, shortly before the Olympic Games, the first core line was designed for 250,000 passengers per day. Nowadays, up to 840,000 daily passengers (DB Netz, n.d.) use the first core line. To handle the flow of passengers, a train runs through the core line every two minutes (two-minute takt). This makes the first core line the busiest in Europe (Schöttle et al., 2023). For a growing city like Munich, the first core line had reached its limit, so a second core line with two more tracks is currently being built. Both core lines run underground through Munich's main stations. This means that the main station needs to be extended and modernized, with the constraint that the infrastructure hub with 450,000 passengers per day operates continuously throughout construction. Besides the underground work, existing buildings will be demolished, and new buildings will be built or updated. Moreover, a precautionary tunnel for a future subway line is to be built. This makes the Munich main station a megaproject within the megaproject of building the second core line. Among many other constraints, this infrastructure project is also very political. So, everything you do is under the microscope of the government, society, and the organization.

With this in mind, in 2021, Annett started helping the Deutsche Bahn project team for the Munich Main Station to better understand the project needs and the related subprojects by developing a paper plan covering 15 years on a monthly basis using sticky notes (see Schöttle and Böker, 2023). After seven workshops in half a year, the plan was finished, and the project team management realized that the current delivery system might not match the project needs and the achievement of the end milestone. The team heard about a new collaborative way of project delivery but didn't have any knowledge about it. That was the starting point of the decision process. Annett convinced the technical project manager to apply CBA to this decision, and because he and the team trusted her, it was a smooth implementation. So, the team came together with a core group to first discover which decision questions needed to be answered. The team quickly aligned on the following two questions:

1. Should the project be delivered with DBB or IPD?
2. Which work scope should be part of the IPD (multiparty agreement)?

Using CBA helped the team to create a shared understanding of why to choose IPD as well as a shared understanding of which scope should be part of the multiparty agreement in a short time. Besides that, including different perspectives helped them to better understand the project constraints. The team was able to align the different perspectives and agreed on the decision outcome. They could organize their argumentation logically and transparently for the management board to argue the use of IPD and the scopes to be included in the multi-party agreement. Most importantly, it created trust and confidence in the decision-making process within the team. Thus, if you want to push a strategic decision through the organization, you need conviction, unity, knowledge, and good documentation and communication (Schöttle et al., 2023). CBA cannot help you with everything, but it does help you to understand knowledge and information gaps, align different perspectives, and make them visible.

The tendering procedure in public projects

Based on the delivery system, the tendering procedure differs, plus there is a difference in how public and private owners design the tendering procedure (Schöttle and Behme, 2024). Where private owners are free to choose the partners they want, public owners must follow specific regulations when tendering a project. These regulations expressed in the law of most countries can be summarized as: (1) select the most economically advantageous tender, (2) the awarding criteria have to be defined precisely to allow a transparent and effective review of the submitted proposals and published before requesting for proposals (verifiable and objective), and (3) the awarding criteria have to be linked to the contract subject, meaning that the contracting authority does not have unrestricted freedom of choice. Furthermore, the contracting authority is free to choose the method of evaluating the submitted proposal information, but still with the exact requirements as the awarding criteria; it must be defined and published when requesting proposals.

Usually, only the price is used to differentiate between the proposals. Why is this mainly the case? It is a simple way to avoid complaints against the procedure as it is relatively easy to describe it explicitly as a number. When using other criteria, they are often weighted lower than the price. So, the public owner asks suppliers to price a project. Then, the public owner usually picks the company that promises to build it cheaper, and that is how we keep "transparency" for public resources spending; thus, the public can see how much money was spent but not necessarily understand the rationale of the decisions and the trade-offs the decision-makers made. Such a method of decision-making, also known as a tendering procedure by "lowest bid," tends to create an unhealthy price competition, resulting in a working environment where bidders hide knowledge and information in order to make claims to get money out of the project to survive after the bid has been awarded (Cox and Thompson, 1997; Darrington and Howell, 2010; Schöttle and Gehbauer, 2013; Fulford and Standing, 2014). This leads to waste in terms of conflict resolution and legal procedures and little value creation for the public resources spent.

We recommend highly analyzing value and cost separately by using CBA, as explained in Chapter 3. Furthermore, a fair and trustworthy procurement process is needed to select the best team to build collaboration. If not, the bidder will go against the owner to maximize the outcome of their strategy. Understanding the owner's value proposition and performance expectations at a very early project stage will also help a team be quicker in the design, and especially in the execution phase.

With the relative weighting of the award factor, the contracting authority thinks it makes its priorities ultimately clear to the bidders. It is an assumption that weighting factors are a powerful tool for achieving the award objective. As explained in Chapter 7, weighting factors is an unsound method. This assumption contains two major mistakes. First, assigning a numerical weight does not make a decision method objective, and, second, the importance of a factor cannot be represented by one numerical point; it is an assessment of an assumption (Suhr, 1999).

Tendering rules and regulations were developed to treat bidders fairly and avoid favoritism. Nevertheless, in doing so, the owner's objectives are often ignored. An important part of public tendering processes is to firmly establish the owner's objectives in the project charter and select a team for whom achieving the owner's objectives is extremely important. But this cannot be done by getting a price based on the specifications. Furthermore, the idea of a fair tendering procedure is right and good, *but no tender process will ever be bulletproof*. On the one hand if public owners want to choose their preferred party, they will find a way to design the tender accordingly. On the other hand, it can also easily happen that a bidder gets chosen that is not beneficial for the project.

Changing the tendering procedure in a public organization

The following case will give you an example of a public owner who has already changed the tendering procedure to avoid a selection mainly based on price.

As a PhD student, Annett went to San Francisco and collected data at the University of California, San Francisco (UCSF) in 2014. In the past, what we usually read and see in the news also applied to UCSF. Projects were delivered with overrun schedules, budgets, and disputes among the different parties. With the high settlement payments, the high construction demand, and the labor shortage in the Bay Area, something needed to change for UCSF to deliver projects successfully. These constraints demanded change, but nothing would evolve without leadership intervention. With a new leadership team in the project department who understood the need for change and a team motivated to initiate the change, the transformation started with the team requesting the state of California to provide an opportunity to change the current, permitted project delivery system using a novel pilot program.

In the pilot program, UCSF changed the way it delivered projects from the traditional design-bid-build (including lowest bid) to a more collaborative delivery system (including a multi-criteria and multi-stage tendering procedure) based on Best Value Selection (Senate Bill 667 (2)). At that time the implementation of IPD was not possible, because due to the law UCSF was not allowed to use a multiparty contract. However, UCSF outperformed in its construction projects compared to other University of California campuses (Schöttle, 2022). That was exciting to Annett as such a comparison in one organization (the University of California) with independent business units did not previously exist. Often, people talk about the uniqueness of a construction project with a focus on the final product instead of seeing the similarities in processes and delivery systems, which have a tremendous impact on projects. According to Lewin (2012), UCSF was a "sociocultural island" that delivered projects on time and within budget. So, what made UCSF so successful? Beside other reasons, Annett learned that the owner consistently ensures it understands its own and the user's needs first to better communicate expectations and value to the project delivery team. This ensured that directions were clear, and that the owner was not disrupting and delaying the project by not being able to decide.

For example, at the UCSF Mission Hall office building project, the owner set up a team that included external help to better understand their own and the user's needs before starting the tendering procedure. This was important to give a clear concept: (1) to the users, since they needed user approval to proceed with the project, and (2) to the future design-and-build team. This building represented a significant change for the UCSF organization itself as some of the medical doctors previously had up to two offices, and with the new building, they changed to a shared desk. This was not an easy transition. So, the owner needed to be prepared to explain why organizational change was critical in its interface with the building's different user groups (see Chapter 2).

Understanding its own needs and value resulted in the owner's clarity during tendering (Schöttle, 2022). The owner made his performance expectations regarding the building and project delivery system clear by articulating these during the competition phase. He left no room for speculation. This is also a perfect example of taking time to better understand and align on purpose, expectations, and value to make decisions faster.

The project-based value was based on specifications regarding the aesthetic integration of the building within the campus, the functionality of the building, and the user experience, by creating a learning environment in which people can interact with each other and be innovative. The value of each building is aligned with the purpose of UCSF, which is to advance health worldwide through research, education, and excellent healthcare. By changing the tendering procedure, the owner spent a lot of upfront time developing a document that resulted in a unanimous decision approving the project by the end users, as well as a phased tendering procedure that reduced the number of bidders to three teams participating in the competition phase. For example, this document consisted of the design and detailed technical specifications.

To promote the integrative approach, all requirements were defined by element or system, not discipline. To understand a product, you had to understand the system, which resulted in a better overall understanding of the building by the bidding teams. After the competition phase, the bidding teams had to submit their technical proposal and a price proposal. The technical proposal was then scored by assigning 1/3, 2/3, or 3/3 to fulfill the criteria. The owner used this score to show the differences between the proposal and not split hairs (Schöttle et al., 2015). *The price proposal was opened after evaluation of the technical proposal to avoid price biases.*

In total, about 260 performance criteria were scored. After assigning the scores, the price proposal was opened, and the ratio price per score was calculated to rank the proposals. We can see how much effort the owner invested in learning about their own needs, the users' needs, and their expectations for the project. Although it was a huge and significant change, the tender evaluation contained two pitfalls: (1) A ratio at the end showing the outcome is not transparent, as a third party will not understand how you get to the ratio and how the scores were assigned. (2) The evaluation with 1/3, 2/3, and 3/3 over- and underestimates the differences between the proposals. Attributes are scored in a way that does not provide a true measure of the benefit in comparison to the other alternatives. There is no understanding of how important specific differences are. For example, it could be that in one factor there is a higher value between the difference of an alternative scored 1/3 and an alternative scored 2/3 than between the alternative scored with 2/3 to the alternative scores with 3/3. By scoring the difference with the same distance of importance, we underestimate the value between 1/3 and 2/3 and overestimate the value between 2/3 and 3/3. The assumption of linearity often comes with scoring systems and leads to outcomes that do not represent the actual value (see Chapter 7). As we learned in Chapters 4 and 7, we usually assume linearity, but preferences are seldom linear. So why do we still hold on to linear assumptions when they clearly make no sense? It is an issue of what we systematically learn in school, at universities, and at work without noticing that this is not helping us in representing our needs. The positive thing about the scoring was that the owner linked the scoring of the alternatives to each other and did not score one proposal after the other. Nevertheless, scoring so many attributes can easily lead to confusion as it is not traceable what the difference between the proposals really is.

Not understanding a ranking can easily lead to complaints about the proposal outcome. In this case, the owner stipulated the price, so there was no ranking regarding the ratio price per score. It was an evaluation based on the fulfillment of factors in each proposal that embodied the owner's objectives and the relative level of performance. The evaluation of the proposals was an intense process, and involved engineering discipline experts. The evaluation team scored all three proposals in comparison to each other.

Now let us assume that the owner did not fix the price and every bidding team had to submit a price proposal. Furthermore, the proposals would be ranked using the ratio price per score. What would have happened? The ratio doesn't help us to understand the ranking of the bidders, when different price proposals are submitted. It is impossible to see why a bidder achieved a particular ratio. Therefore, a transparent comparison of the alternatives on beneficial differences will help to understand the outcome of a procurement process, so the likelihood of claiming against a proposal will be less. Furthermore, the tendering procedure could be gamed, as *the ratio might attract speculative bidder behavior*. Bidding teams can easily set up speculative price proposals to win the bid. You could win the bid with a poor technical proposal and a low speculative price (Schöttle and Arroyo, 2017). The owner also cannot ask whether it is worth paying more for a more advantageous alternative.

UCSF started to change and step by step they continuously improved the way they delivered projects. Nowadays, they are also allowed to choose IPD to deliver projects and as such, UCSF is currently building a $4.3 billion hospital complex under a multiparty contract with 14 parties.

The CBA Tabular method with the Prior Anchoring Process (PAP)

As mentioned in the previous section, the public tendering procedure is based on regulations to guarantee a fair procedure. This means the publication of the tendering documents needs to include how proposals will be evaluated. For most countries, this means it is not enough to state that the proposals will be assessed based on CBA. It needs to be shown in detail how IoA are assigned. Therefore, Suhr (1999) introduced the prior anchoring process (PAP) as an add-on to the CBA Tabular method for the public tendering procedure. The PAP consists of five steps:

1. Define the factors.
2. Define the must-have and want-to-have criteria for each factor.
3. Estimate the expected range of attributes for each factor.
4. Decide the expected range of advantages in each factor by defining the minimum requirement attribute and the best expected attribute as anchors to identify future advantages. Calculate the anchor-statement advantage, which is the difference between the minimum requirement and the best expected attribute.
5. Decide the importance of the expected advantages by selecting the paramount advantage among the anchor-statement advantages. Then, assign the importance weight of each anchor-statement advantage compared to the paramount advantage, and for each factor, construct either a preference curve for quantitative data or a preference chart for qualitative data.

These definitions and the statement that the proposals will be evaluated using CBA must be published in the tendering documents to ensure transparency for all bidders. After receiving the proposals, the owner evaluates them based on the PAP (see Figure 8.2). To avoid biases, the proposal should also be evaluated anonymously, and the price proposal needs to be submitted separately from the value proposal. This might be difficult for proposals that include a competition process, as it will be very easy to identify the proposal with the bidding team. Here, a diverse

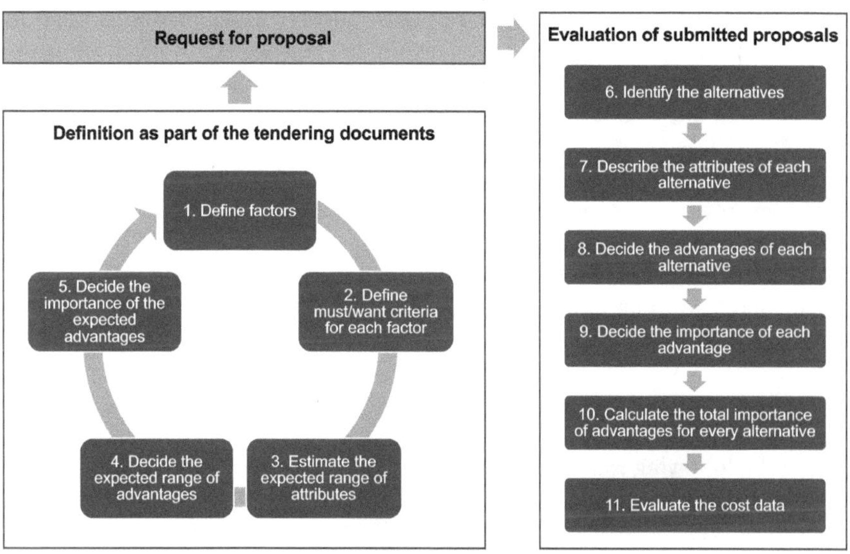

Figure 8.2 Process steps of CBA in the tendering procedure
Source: Schöttle and Behme (2024).

team representing the architecture, technical systems, structure, landscaping, sustainability, facility management, user, and owner perspective must take care of the evaluation to avoid favoritism (see Chapter 2). Setting up a PAP gets the owner to reflect on what they want, but it is also a big effort to define everything in detail, especially on large projects. In several countries it might even be necessary to define the whole preference curve/chart upfront to comply with the law. This would add even more work on the owner's side and might lead to owners not wanting to change from a scoring system to CBA. Therefore, it would be advantageous if the rules and regulations were adjusted to be able to use CBA and really investigate the differences between proposals and not force public owners to use unsound methods (Suhr, 1999; Schöttle and Behme, 2024).

The prior assessment of the minimum requirement attribute, the best expected attribute, and the resulting anchor-statement advantage can be difficult in the case of qualitative data because of the higher level of interpretation (Schöttle and Behme, 2024). To avoid the interpretation, the preference chart must contain a range of expected advantages; therefore, an expected range of attributes needs to be defined. Suhr (1999) stated that preference curves are only for quantitative data and that qualitative data has to be presented using preference charts. We also think qualitative data could be presented using preference curves to make it easier for practitioners to apply. Nevertheless, it is suggested that, in this case, a set of preference curves should be developed from which the team can choose in order to avoid an abstract mathematical discussion.

Furthermore, as stated above, the criteria must be clearly defined to avoid interpretation during the evaluation process. Another tricky point is the relation between price and value, which usually must be represented in relation to each other (see Chapters 4 and 5). This doesn't restrict the owner from using the IoA vs. price diagram to show the relation, but it might be challenging to state why an alternative is preferred in comparison to another. For example, if two alternatives are close to each other, how do you choose between one that is less valuable but lower in price, while the other contains a higher value but is also more expensive? This is a big dilemma of public sector contracting. The reasons for choosing one over the other must be transparent and traceable to avoid conflicting situations where bidders might complain about the procedure. This could lead to an infinity of rules defined to deal with trade-offs (Schöttle and Behme, 2024).

Example of the Prior Anchoring Process

Using the following example, we want to explain the PAP more clearly. Let us assume you are in the position of a public owner who wants an office building interior renovation. You have already decided that you want to deliver the project using DB. Now, you want to select the team to design and build based on the conceptual design description. The pre-selection phase is already over, and you have selected two teams for the competition phase. For the evaluation, you have prepared the CBA Tabular method showing the PAP so that both teams know how their proposal will be assessed. To keep this example as simple as possible, we will only use three factors: light system, workspace approach, and replacing HVAC air filters.

For the first factor, "Light system," you defined your criterion as "less kWh/m^2/year, is better." Then, you set the minimum requirement for the criterion. In this example, we choose 104 kWh/m^2/year. This attribute defines your reference points of the preference curve on the left side. It also means that an alternative, for example, 110 kWh/m^2/year, is unacceptable. Thus, if a proposal contains a number higher than 104 kWh/m^2/year, it must be excluded from the evaluation process. Then, you estimate the best expected attribute a proposal might contain. In this example, you estimated 41 kWh/m^2/year. To calculate the anchor-statement advantage, you then build the difference between the minimum requirement and the best expected attributes. Here, it is 63 kWh/m^2/year (see Table 8.1).

You continue with this procedure for all factors. Now, where you determine the anchor-statement advantage within all factors, you select the most important advantage from the anchor-statement advantage. This will be your paramount advantage (see also Chapter 5 for CBA method). For this case, we choose an IoA scale from 0 to 100. Therefore, we assign a score of 100 to our paramount advantage which in this example is 63 kWh/m^2/year less. Based on the paramount advantage, you follow the already known procedure and weight the other anchor-statement advantage.

Table 8.1 Example of the CBA Tabular method in the tendering procedure

Factor (Criterion)	*MR attribute*	*BE attribute ASA*	*Weight of ASA*	*Preference curve/chart*
Light systems	104 kWh/m^2/year	41 kWh/m^2/year		see Figure 8.3
(The less kWh/m^2/year, the better.)		63 kWh/m^2/year	100	
Workspace approach	Standardized rooms with mobile separating walls	Completely flexible space, adjustable for every situation (focus, group and workshop meetings)		see Figure 8.4
(The more flexible, the better.)		Significantly more flexibility regarding space	80	
Replacing HVAC air filters	Every 3 months	Every 6 months		see Figure 8.5
(The less often the exchange is necessary per year, the better.)		Two times less per year.	20	
Total IoAs			**200**	

As stated previously, the way in which the proposal will be evaluated must be clearly defined, and thus, the preference curves or charts need to be included. As a result, we have to define the preference curve/chart for every factor. To do so, it is recommended to think about further attributes and how their advantages might fall into a factor's preference curve/chart. Figure 8.3 presents the preference of the factor "Light system."

For qualitative data, we need to draw the preference chart (see Figure 8.4). Figure 8.4 shows the chart of the factor "workspace approach." Defining

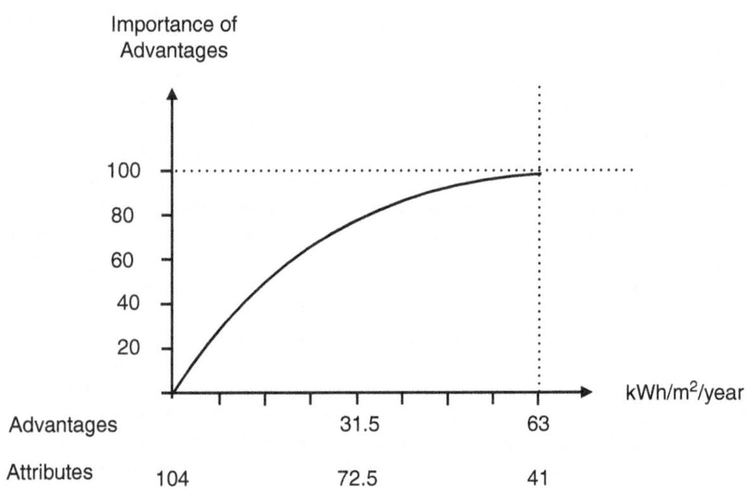

Figure 8.3 Preference curve of the factor "light system"

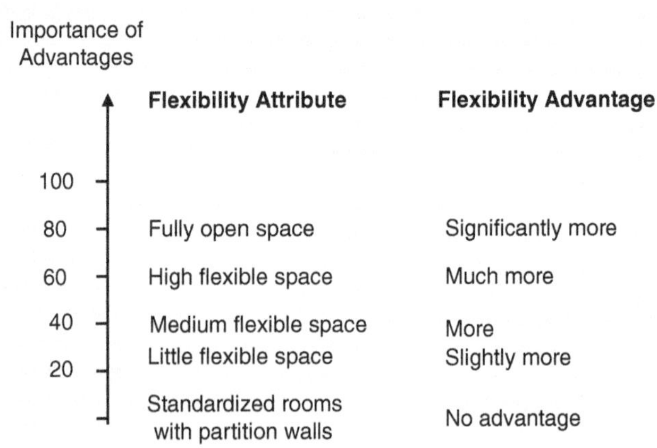

Figure 8.4 Preference chart of the factor "workspace approach"

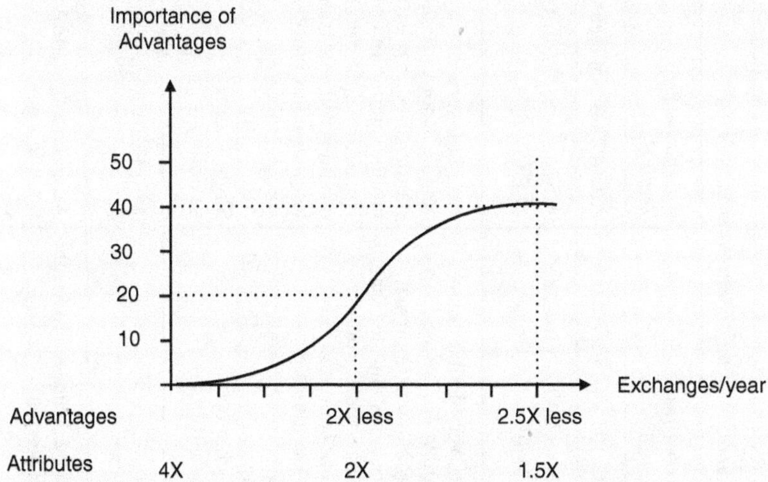

Figure 8.5 Preference curve of the factor "replacing HVAC air filters"

the preference chart can be tricky as you must insert a range of attributes/advantages to be clear with the evaluation later. It is more difficult as you can't just follow a curve as you can with quantitative data. It sometimes leaves gaps that can lead to interpretations and discussions when evaluating. This often causes contract managers to fear, but think about the UCSF case. They changed the evaluation method during the assessment of the proposals to make the differences clear, and the bidding teams were okay with it. Rather than assuming things, open communication about the change helped.

The third factor, "replacing HVAC air filters," in our small example, shows that an alternative can also exceed the weight of the anchor-statement advantage. In this case, the preference curve must be extended (see Figure 8.5). Therefore, considering extension when defining the preference curve is essential.

After defining the preference curves/charts, the table together with a description of the procedure is ready to be part of the tendering document. After the submission deadline, the proposals can then be assessed based on the PAP by following step-by-step the CBA Tabular method by first identifying the proposals and describing the attributes of each alternative. Then defining the advantages and the importance of the advantages based on the predefined preference curves and charts (see Table 8.2).

Note that within the current law, the process cannot add new information to differentiate between the teams, although this might occur during the competition phase as new perspectives and information are exchanged and collected. As the small example shows, defining the preference curves/charts takes time and can be especially difficult for the qualitative data.

Table 8.2 Assessed proposals from the example using the CBA Tabular method

Before request for proposal					*Evaluation of the submitted proposals*			
Factor (Criterion)	*MR attribute*	*BE attribute ASA*	*Weight of ASA*	*Preference curve/chart*	*Alternative 1: Team A*		*Alternative 2: Team B*	
Light systems	104 kWh/m²/year	41 kWh/m²/year		see Figure 8.3	Att.: 101 kWh/m²/year		Att.: 49 kWh/m²/year	
(The less kWh/m²/year, the better.)		**63 kWh/m²/year**	100		Adv.: 3 kWh/m²/year	IoA: 8	Adv.: 55 kWh/m²/year	IoA: 96
Workspace approach	Standardized rooms with mobile separating walls	Completely flexible space, adjustable for every situation (focus, group and workshop meetings)		see Figure 8.4	Att.: Very flexible. Spaces (hard walls) are used in a very modular approach to be easily adjusted for changes. Focus rooms can be converted into huddle rooms, etc.		Att.: Not very flexible. Room sizes are standardized, but with mobile separating walls. Irregular neighborhood modules will constrain long-term flexibility.	
(The more flexible, the better.)		**Significantly more flexibility regarding space**	80		Adv.: Much more flexibility	IoA: 70	Adv.: Slightly more flexibility	IoA: 25
Replacing HVAC air filters	Every 3 months	Every 6 months		see Figure 8.5	Att.: Every 8 months		Att.: Every 6 months	
(The less often the exchange is necessary per year, the better.)		Two times less per year.	20		**Adv.: 2.5 times less per year**	IoA: 40	Adv.: Two times less per year.	IoA: 20
Total IoAs			**200**			**118**		**141**

Table 8.3 Recommendation for the tendering procedure

	Current process	*Recommendation*
Approach	Project needs to be defined 100% up-front and it is often assumed that every bidder delivers the product with the same quality	With 100% likelihood there will be changes, so the selection needs team criteria to differentiate between the bidders
Process	Mostly one-step process sometimes with pre-qualification	Multi-stage process
Decision criteria	Mainly price, and if other factors are used, they are often percentage and not decisive	Price is the only information for the trade-off
Method	WRC/BVS	CBA

Source: Schöttle (2022).

Reflection on the prior anchoring process

What needs to happen? The tendering procedure needs to change to highlight the differences between proposals and to install a fair process, see the recommendations in Table 8.3. Therefore, owners must be empowered and enabled to make the awarding decision proactively. It should be noted that the current system creates a reactive process because, after the request for proposal, the client must follow the tender document and cannot improve it. In line with Schöttle and Behme (2024), we suggest that the process should be actively designed. This also includes giving the owner responsibility for the awarding process. Ultimately, corruption in procurement can only be counteracted through transparency, a diverse selection team, and traceability. By doing so, public authorities will put themselves in a proactive position based on getting more value through collaboration instead of a selection based on risk reduction and fear.

Key points

- Tendering is the first interaction between an owner and the potential partner and the delivery system defines the tendering, therefore, the decision of how to deliver the project is crucial and should be well thought out and here CBA can help.
- Owners must invest in the extensive up-front work required to establish a stable tendering procedure.
- CBA can create a solid and transparent argumentation to justify the why of a decision through an organization.

- If the price is not set beforehand, the price proposal should be opened after the evaluation of the value proposal to avoid biases regarding the price.
- Ratios of value and price support speculative bidder behavior and therefore need to be avoided.
- The PAP process helps to be more transparent upfront with decision-making. However, the point is not to create rules for every possible trade-off.
- To act proactively, regulations and rules for tendering procedures need to change to be able to discover and compare the proposals.
- No tendering rules and regulations will ever be bulletproof. Therefore, a diverse team representing the different perspectives must take care of the evaluation to avoid favoritism.

References

Ballard, G. (2012). “Should Project Budget Be Based on Worth or Cost?” In *Proceedings of the 20th Annual Conference of the International Group for Lean Construction.* San Diego, USA.

Carpenter, N. and Bausman, D.C. (2016). “Project Delivery Method Performance for Public School Construction: Design-Bid-Build Versus CM at Risk.” *Journal of Construction Engineering and Management*, 142(10), 05016009.

Chih, Y.-Y. (2010). “A Decision-Support Framework for Choosing a Project Delivery System (PDS) in a Multi-Project Environment.” PhD thesis, University of California, Berkeley, Available at: https://escholarship.org/uc/item/76j9q5pk.

Cox, A., and Thompson, I. (1997). “ ‘Fit for Purpose’ Contractual Relations: Determining a Theoretical Framework for Construction Projects.” *European Journal of Purchasing and Supply Management*, 3(3), 127–135.

Darrington, J., and Howell, G. (2010). “Can Optimized Projects Require Optimized Incentives?” In *Proceedings of the 18th Annual Meeting of the IGLC*, Haifa, Israel, pp. 591–600.

DB Netz (n.d.). “Die Situation heute.” Available at: https://www.2.stammstrecke-muenchen.de/die-situation-heute.html (accessed February 10, 2023).

Elfving, J.A., Tommelein, I.D., and Ballard, G. (2005). “Consequences of Competitive Bidding in Project-Based Production.” *Journal of Purchasing and Supply Management*, 11(4), 173–181.

Eriksson, P.E., and Laan, A. (2007). “Procurement Effects on Trust and Control in Client-Contractor Relationships.” *Engineering, Construction and Architectural Management,* 14(4), 387–399.

Eriksson, P.E., and Westerberg, M. (2011). “Effects of Cooperative Procurement Procedures on Construction Project Performance: A Conceptual Framework.” *International Journal of Project Management*, 29(2), 197–208.

Fischer, M., Ashcraft, H.W., Reed, D., and Khanzode, A. (2017). *Integrating Project Delivery*. Hoboken, NJ: Wiley.

Forbes, L.H., and Ahmed, S.M. (2010). *Modern Construction: Lean Project Delivery and Integrated Practices*. Boca Raton, FL: CRC Press.

Fulford, R., and Standing, C. (2014). "Construction Industry Productivity and the Potential for Collaborative Practice." *International Journal of Project Management*, 32(2), 315–326.

Heidemann, A. (2011). *Kooperative Projektabwicklung im Bauwesen unter der Berücksichtigung von Lean-Prinzipien: Ent wicklung eines Lean-Projektabwicklungssystems: Internationale Untersuchungen im Hinblick auf die Umsetzung und Anwendbarkeit in Deutschland.* Karlsruhe: Karlsruher Institut für Technologie.

Heidemann, A., and Gehbauer, F. (2010). "Cooperative Project Delivery in an Environment of Strict Design-Bid-Build Tender Regulations." In *Proceedings of the 18th Annual Meeting of the IGLC*, Haifa, Israel.

Johnson, T.R., Feng, P., Sitzabee, W., and Jernigan, M. (2013). "Federal Acquisition Regulation Applied to Alliancing Contract Practices." *Journal of Construction Engineering and Management*, 139(5), 480–487.

Konchar, M., and Sanvido, V. (1998). "Comparison of U.S. Project Delivery Systems." *Journal of Construction Engineering and Management*, 124(6), 435–444.

Lahdenperä, P. (2012). "Making Sense of the Multi-Party Contractual Arrangements of Project Partnering, Project Alliancing and Integrated Project Delivery." *Construction Management and Economics*, 30(1), 57–79.

Lewin, K. (2012). *Feldtheorie in den Sozialwissenschaften: Ausgewählte theoretische Schriften*. Bern: Huber.

Love, P., Davis, P., Chevis, R., and Edwards, D. (2011). "Risk/Reward Compensation Model for Civil Engineering Infrastructure Alliance Projects." *Journal of Construction Engineering and Management*, 137(2), 127–136.

Rooke, J., Seymour, D., and Fellows, R. (2004). "Planning for Claims: An Ethnography of Industry Culture." *Construction Management and Economics*, 22(6), 655–662.

Rosenfeld, Y. (2014). "Root-Cause Analysis of Construction-Cost Overruns." *Journal of Construction Engineering and Management*, 140(1), 04013039.

Schöttle, A. (2022). *Voraussetzungen und Anreize für die kollaborative Lean-Projektabwicklung im öffentlichen Bausektor: Stärkung der Autonomie, Kompetenz und Verbundenheit von Projektteams.* [Doctoral dissertation, Karlsruher Institut für Technologie]. doi.org/10.5445/IR/1000152023

Schöttle, A., and Arroyo, P. (2017). "Comparison of Weighting-Rating-Calculating, Best Value, and Choosing by Advantages for Bidder Selection." *Journal of Construction and Engineering Management*, 143(8). doi.org/10.1061/(ASCE)CO.1943-7862.0001342

Schöttle, A., Arroyo, P., and Bade, M. (2015). "Comparing Three Methods in the Tendering Procedure to Select the Project Team." In *Proceedings of the 23rd Annual Conference of the International Group for Lean Construction*, pp. 267–276,

Schöttle, A., and Behme, F. (2024). "A Critical Analysis of Choosing By Advantages Implementation in the Tendering Procedure Based on EU Directive and German Legislation." In *Proceedings of the 32nd Annual Conference of the International Group for Lean Construction*, pp. 1244–1254. doi.org/10.24928/2024/0223

Schöttle, A. and Böker, A. (2023). "Integrating a Strategic Milestone and Phase Plan (SMPP) as a New Level of the Last Planner System (LPS): An Investigation on Megaprojects." In *Proceedings of the 31st Annual Conference of the International Group for Lean Construction (IGLC31)*, pp. 1418–1427. doi.org/10.24928/2023/0252

Schöttle, A., and Gehbauer, F. (2013). "Incentive Structure in Public Design-Bid-Build Tendering and Its Effects on Projects." In *Proceedings of the 21st Annual Conference of the International Group for Lean Construction*, Fortaleza, Brazil.

Schöttle, A., Guerreiro, R., Haubrich, A., Ditandy, M., and Behme, F. (2023). "Organizing the Argumentation for Changing the Delivery System Using Choosing by Advantages (CBA)." In *Proceedings of the 31st Annual Conference of the International Group for Lean Construction (IGLC31)*, pp. 990–1001. doi.org/10.24928/2023/0245

Suhr, J. (1999). *The Choosing By Advantages Decisionmaking System.* Westport, CT: Quorum.

Tsao, C.C.Y., Tommelein, I.D., Swanlund, E., and Howell, G.A. (2000). "Case Study for Workstructuring: Installation of Metal Door Frames." In *Proceedings of the 8th Annual Conference of the International Group for Lean Construction*, Brighton, UK, pp. 1–14.

Wardani, M.A.E., Messner, J.I., and Horman, M.J. (2006). "Comparing Procurement Methods for Design-Build Projects." *Journal of Construction Engineering and Management*, 132(3), 230–238.

9 Building a collaborative decision-making process

Until now, we have focused on the final assessment of alternatives and giving guidance on how to decide between two or more alternatives using a multicriteria decision method: Choosing by Advantages (CBA). We have presented these decision-making methods logically and straightforwardly. We also explained how CBA must be applied in the tendering procedure. However, the final assessment is only part of a more comprehensive process, which includes many more sub-processes, and the reality of making the decisions can be complicated. The activities seldom happen as straightforwardly as described in models and theories, and decision-making includes people, power dynamics, feelings, history, identity, background, tradition, agendas, interests, and time pressures.

This chapter offers strategies and examples for building a more comprehensive decision-making process for inspiration. We will tie many of the key points from the previous chapters together in one coherent process. To do this, we introduce some Lean concepts and methods to support efficient decision-making. Lean is a philosophy based on a set of principles and methods that aim to increase value for the client while reducing waste.

With a collaborative decision-making process, you are more likely to stay true to the CBA principles despite the need to adapt to specific and sometimes complicated circumstances. Finally, an example will demonstrate some benefits of implementing CBA with regard to value, time, and cost.

The value of an agreed collaborative decision-making process

According to Schöttle et al. (2018), a decision-making process should consider a decision method, the structure of the process, governance, and documentation of the decision. So far, we have mostly described the decision method and why we recommend CBA. This chapter will expand into the wider process, starting after a decision has been recognized and ending with implementation and assessment of consequences. The following are some of the key activities in a collaborative decision-making process:

DOI: 10.1201/9781003518310-10

1. Defining the decision and its purpose
2. Involving the right people
3. Preparing the team for collaborative decision-making
4. Planning the decision and understanding interdependencies
5. Value and budget management through decision-making
6. Exploring alternatives and factors for innovation
7. Decision-making and documentation
8. Implementation and learning

An agreed approach to the above key activities will provide you and the team with a language and a shared understanding of how the decision-making process should be. This allows you to adapt to specific circumstances while staying true to the CBA principles. Below is our offering of such a process and necessary considerations, built on accumulated knowledge from many projects and decisions and kept at a general level. Many of our experiences stem from major projects but have also been used on smaller projects or in everyday decision-making. You should adapt our general outline of a process to your specific context, and you should specify and integrate the relevant steps into your management systems and practice and make templates and documents suited to your needs. We will start with an example to illustrate the value of having such a clearly defined process for a project.

Decision-making in a hospital project

Annett was impressed by the team's decision-making approach when she studied the UCSF Mission Bay Medical Center project, a $1.5 billion building complex with specialized facilities for children, women, and cancer care, opened in 2015. The decision-making process was not only well-defined, it was also transparent for everyone. A major challenge of the project was to design just enough to get the permit, knowing that the design would later need to be adjusted based on the rapid technological change in the healthcare industry. Another challenge was to keep track of ideas and innovations and decide which should be executed to generate an efficient design phase.

Therefore, the team implemented a decision-making process to make decisions on time and at the lowest responsible level in the organization. The decision-making process consisted of four hierarchical levels: (1) individual cluster group, (2) captain's level, (3) senior leadership principals, and (4) management committee. The cluster group consisted of eight divisions: site, structural, exterior wall, mechanical and plumbing, electrical/low voltage, special systems, interiors, and equipment, and focused on the technical issues. They interacted, discussed, negotiated, and traded targets with each other to make progress and meet the budget.

If ideas turned into alternatives, they were documented and shown to the impacted stakeholders, and input was collected before finalizing the decision.

The organizational level where the decision was anchored depended not only on cost impact but also on the value generation. For example, if a decision impacted the design of the building, schedule, cost, or sustainability, the decision had to be escalated to the senior-level management. Nevertheless, the process helped the team to keep decisions on a lower level. Most decisions were made on the captain's level; only 2% were escalated to senior-level management.

The process also defined the time frame for finalizing decisions at each level so as not to risk delaying a decision. For example, the captain's level had three days to decide, and the senior leadership had one week. The bottom-up approach of quickly checking the possibilities and consequences and making the decision on the right organizational level resulted in people spending less time on ideas that wouldn't be realized, resulting in faster and on-time decisions and higher productivity.

The design process ended successfully. The decision-making process produced $55 million worth of design improvements, adding value to the project and not getting stuck in negative iterations (Schöttle et al., 2018). A well-defined decision-making process must be implemented for big and megaprojects to enable the team to move forward. Teams need to plan how decisions will be made before they face significant decisions (Schöttle, 2022). This doesn't mean that the process cannot be improved during execution, but there needs to be an agreed reference to be able to adjust. Not having a decision-making process will lead to wasting time during delivery. The decision-making process created a high level of trust among the team members, as everybody knew when a decision would be made and what status every decision had.

Defining the decision purpose

If you want a decision made, you should be able to articulate a clear purpose to assess whether time and effort should be used to explore this decision. There are many questions you need to consider, such as: What input do you expect by inviting stakeholders? What is it you are looking to achieve? There could be different purposes, for example, to create a shared understanding of the alternatives and the decision, to gather ideas for improving the alternatives in specific areas, to have an efficient process and quick implementation of the decision, to create a thorough decision brief for politicians; or a combination.

To understand the purpose, you need to understand the decision context, as well as the concerns of the stakeholders. A clear purpose could stem from a vision, a strategy, or a set of imperatives or values stated by the client. In some projects, such a "value key" is defined as a set of factors for which improvements are valued and appreciated. Identifying the needs and preferences of the people impacted by the decision, such as the users, could also give direction for the decision and how it should be made. For example, when UCSF and the design team renovated the Anatomy Teaching Lab at the Parnassus campus, they understood their purpose was to create the best possible learning environment for the students. So, instead of designing it as a sterile environment, as was normal practice and how it was before, they designed it as a colorful and vibrant place to engage students. The owner even attended classes to better understand the needs of the students and professors (Schöttle, 2022).

The purpose of the decision helps define the requirements for the decision-making process and helps determine when decisions should be considered more carefully, whether this process should aim at quick implementation, ensure a thorough risk assessment, result in added value through improvements or innovations.

Involving the right people

The purpose also helps define who should be involved in the decision. We recommend making the decisions at the lowest responsible level in the organization while considering who has the knowledge, skills, and ability to imagine the consequences of the decision, as in the hospital example above. Having an agreed governance structure will allow decisions to be made at the right level without placing unnecessary burdens on the higher level of the organization. Decisions often escalate unnecessarily due to a lack of clarity, making the higher management level a bottleneck for decision-making. Delegating authority could ease the burden of these management levels and allow management to focus on more strategic decisions and forward planning.

Agreeing on roles and responsibilities is necessary to ensure that expectations for all participants are clear so the process will be efficient (Christensen et al., 2023). Generally, each decision should have an owner accountable for the decision process and the actions to implement the decision. The owner only participates in all meetings and workshops if she has a clear contribution. The owner is also responsible for ensuring the decision has a clear purpose.

Involving the right people is crucial to getting all relevant voices heard (see Chapter 2), but assessing who to involve and who to inform can be difficult. Traditionally, people have argued their case by emphasizing the importance of their discipline. No one wants to be unimportant in decision-making, and the fear of missing out often leads to crowded rooms and inefficient meetings and decisions.

Two main groups need to be involved: those with knowledge and those with influence. You will need those who can contribute professional knowledge of alternatives, factors, criteria, and information on attributes. We can call them specialists. They are experts with specific knowledge or with insights of certain stakeholders. You also need those with significant influence and responsibility for the outcome and implementation of the decision. It can be managers, authorities, or other stakeholders you depend on to approve and implement the final decision.

It is okay to leave out those who have only marginal input or invite people for parts of the workshop. Determining the decisive factors (factors that differentiate the alternatives for the given decision context) and the alternatives is an iterative process. Hence, the number of people and disciplines included in decision-making can change in different iterations, and the attendees could be a core group and additional participants at different stages. Making this process transparent and explaining reasons for different engagement levels is important to foster collaboration and trust in the decision outcome.

Decisions spanning several organizational units, hierarchical levels, or disciplines should be led by a facilitator. In Chapter 2, we described how everyone is biased and has preferences that impact our actions. A facilitator must not have a stake in the decision and can, therefore, focus on the process, ensure progress, ask questions to move the team forward, foster psychological safety, and navigate the biases of those involved in decision-making. The owner should, therefore, hand over the responsibility of the decision process to the facilitator. The facilitator must know the decision-making methods and have the personal competencies to facilitate and prepare the group for collaborative decision-making.

Preparing the team for collaborative decision-making

If the team has never used CBA before, training will be needed to ensure trust in the process and that participants know what is expected of them. Making decisions using CBA is a change for most people and could lead to confusion, tension, or nervousness. For some, it can be seen as a challenge to their authority, and for others, it can be nerve-racking to speak up and discuss things in front of senior managers unless in a psychologically safe environment.

In one case, Randi experienced a senior manager leaving the decision workshop before time after having argued that he did not trust the process and, therefore, was unwilling to continue. The workshop was paused, and the remaining part was replanned a week later. In the following week, Randi had several discussions with the manager about the purpose of the CBA Tabular method and the specific decision. The manager revealed that he had not been prepared for the change in responsibilities, and therefore, he could not argue his case in the first workshop and felt helpless. His emotions inhibited him from participating constructively. After thorough preparation, he joined the last part of the workshop, and subsequently, he became one of the greatest advocates for CBA and has applied it in more projects on his initiatives.

Change can be difficult, and when the pressure is high, we know we are more likely to turn to System 1 thinking (Kahneman, 2011) and just jump to conclusions without checking if we have the right information. Therefore, we advise that anyone wanting to implement CBA in their project or organization should also consider preparing the team for collaboration using CBA. If the team receives the right training and onboarding, they are more likely to be true to the intentions behind the methods, even under pressure.

Participants in decision processes need to understand the vocabulary, the processes, and principles. They need to know what is expected of them and how to contribute at the right time at the right level. Facilitators need to be masters, meaning they need to know intuitively what the intention of the methods and principles is in order to be able to adapt rightfully when needed (Dreyfus and Dreyfus, 1980).

To learn CBA and become competent, you as an individual could consider following this strategy, according to Jim Suhr (1999):

- Step 1: Learn, verify, and begin using just one small set of definitions, principles, and methods (concepts and methods) of sound decision-making at a time. This is the key—just one small set at a time.
- Step 2: Unlearn (learn not to use) the corresponding unsound concepts and methods.
- Step 3: Relearn and relearn (many times) the concepts and methods that you learned in step 1.
- Step 4: Practice and practice (many times) the concepts and methods that you relearned in step 3. Practice them until you are applying them automatically and unconsciously. An intellectual understanding of how to make sound decisions is not enough. It will not produce a change in behavior. You must internalize how to make sound decisions.
- Step 5: Teach others sound decision-making definitions, principles, and methods. (While the first four steps are vital, this step is optional. But we highly recommend it for those with adequate teaching skills.)

Depending on the needs of the learners and the purpose of the decision at hand, different types of training could be offered. Would you need to train a group to make a critical decision quickly? Would you like to train a group of facilitators to support decision-making in the long term? You also need to consider different learning styles for different people. If you are new to CBA, support from an experienced coach might be helpful in any case, and the coach could help you decide on the right path, which could include the following types of training or a mix (Arroyo et al., 2019).

- *Coaching several decisions*: Engage an external coach to facilitate and coach several decisions. After making several decisions (4–20) using CBA, practitioners will have learned the practice and have had the chance to integrate that knowledge, leading to a greater understanding of why CBA focuses specifically on the importance of advantages involved in decisions. Practitioners will begin to take initiative in the decision-making process by, for example, identifying alternatives without being prompted by the coach, investigating factors and criteria, and coming to meetings with a prepared list of attributes. In most cases, by the end of the fourth to the sixth decision made with the CBA coach, practitioners will have developed skills that qualify them as "advanced beginners" to "competent," but without necessarily having fully understood the background or being able to bring this knowledge forward to others.

- *One-day or two-day workshops*: By engaging with a coach to conduct a one-day or two-day training workshop, you can gain in-depth knowledge of the principles and methods, but in an artificial setting. The participants will recognize the benefits of the methods and learn about part of the theory. However, the skills are not yet routine, and the participants will find it difficult to apply them in practice when conflicting approaches are involved, and would still need an experienced coach to facilitate the decision-workshops.
- *Formal training*: This could be 12 weekly 2-hour sessions conducted remotely or in person. This extended training format allows for in-depth study of CBA concepts, vocabulary, principles, and methods. Participants read the relevant literature on the topic and join learning exercises and homework tasks. The structure allows participants to think through different application contexts, ask questions, and learn from their peers as well as from their coaches. After completing the training, students' knowledge of CBA varies from "advanced beginner" to "proficient." However, becoming a skilled facilitator would require that the training includes facilitation of actual decisions with support of a coach as it is as much a craft as it is a theory.

Having taught CBA many times, we have found that the best way to learn CBA if you want to become a facilitator is a combination of learning by doing with the support of an experienced coach supplemented with training in theories and principles. Without this support from a coach, newcomers will likely fall back on traditional decision-making or are less likely to understand the decision-making process.

Planning for decision-making and understanding interdependencies

Efficient decision-making requires planning, just like all other deliverables and activities. How often have you waited to know the outcome of a decision? Perhaps you have been chasing the person you expected to make the decision. Or you have progressed, expecting a certain decision outcome and then realized you would have to redo your work, as the decision did not end with the expected outcome? As decision-making is crucial for working efficiently, we should include the key activities of the decision process in the main schedule. This creates transparency and shows when it is possible to contribute to the decision and when you can expect an outcome.

The Last Planner System™ (Ballard 2000) is a well-known Lean construction management practice that has also been shown to support CBA particularly well. Here, the basic principle is to involve the people doing the work (the last planners) in planning their own work instead of mindlessly following a schedule made by managers or planners based on assumptions. The system also allows everyone to get support to ensure they have all the necessary prerequisites for their upcoming tasks and progress as planned (make work ready for execution). This process of making tasks ready for execution can reveal that certain decisions are needed, which can trigger a decision-making process that needs attention in the mid-term.

In the long term, the overall plan is made in collaboration using a pull mechanism, which means the team works backward from the end milestone to start a project phase. When doing this, the team can identify when major project decisions must be made and plan accordingly. For example, you need to decide the alignment of a road before you decide whether any crossovers are needed and what design these might have. When decision-makers plan decisions in a logical sequence, thinking about who needs what information by when (pull mechanism), they reduce the amount of negative iteration of the decisions.

Understanding the relationship between different decisions is not always simple. Remember, as we stated in Chapter 1, making decisions is about predicting the consequences for the future. Many times, decisions are interdependent. For example, the road's alignment impacts the crossovers' design, but the crossovers might also influence the alignment. Furthermore, if a decision changes, it can affect multiple other decisions. This can easily throw a project's schedule and budget out of control, triggering unprepared teams to start looking for a scapegoat while defending their interests.

Including the right people early in the decision-making process, planning decisions, and making interdependencies among decisions as transparent as possible will help the team be informed and prepared. This will give you an overview of the main decisions and allow you to prioritize which decisions to spend more time and resources on and avoid negative iterations, as described in Chapter 2.

Value and budget management through decision-making

Decisions impact the budget of the project. As mentioned in Chapter 3, we want to avoid a hyper-focus on cost and use some Lean principles. For example, we want to allow money to move across organizational boundaries (e.g., from one building system to another or from one department to the next), ensuring the best value for money. But to do this, we need to understand what value means and how we meet the same criteria for lower costs.

Target Value Delivery (TVD) is a Lean method that seeks to achieve a target value, a desired performance outcome for the whole building or project, at a certain cost set in agreement with the owner/client, which is typically attractive to all the parties and requires innovation to steer to both target value and cost. This method contrasts with the traditional way, in which the design is completed first, and then the cost of the building/project is determined according to market conditions. Here, we will not explain in detail how to implement TVD, but we will highlight that functional teams, often working on a building system or project function, will collaborate to create design alternatives and choose an alternative that fulfills a desired performance at a target cost that is below the current budget, or they can decide to go above the current budget if the entire project value will be better by doing so. This means that the extra cost is funded by another building system or department that can produce the expected value and reduce its cost. This process is key to fostering innovation and system thinking to optimize the whole and not just each individual part. Maintaining a decision log with ideas to explore and evaluate across disciplines is key.

Exploring alternatives and factors

Defining factors, criteria, and alternatives will always be an iterative process. It can be challenging to manage for the facilitator, but, if done well, it can allow for innovative thinking and improvements. No normative advice can be given here that ensures you will always succeed, but we will offer some general advice to consider to navigate this phase.

The quality of the decision relies on many things, including the quality of the alternatives. In Lean management, the use of set-based design is encouraged for innovation. Set-based design is a design method that encourages the exploration of alternative approaches for design solutions and carries them along parallel paths, gradually eliminating them until the preferred approach is chosen (Ward et al., 1995). This prevents the traditional point-based approach, which is the premature selection of an alternative that, in the end, results in re-work or subpar value to the client when the team finds too late that it is not the optimal choice (Ward et al., 1995). A simple example of the use of these concepts is as follows: imagine you are organizing a meeting, and you select a time that suits you, then you go back and forth with every person to check if they have time on that specific day and time; if one person cannot come, you choose another time and start the back and forth again, and again, this is a point-based approach. If you instead send out a poll with all your options, people can assess when they have time and give you all their available slots; you will get your answer quickly based on the set-based approach.

Having a clear purpose of the decision will help define factors, but all aspects where significant differences exist between the alternatives should be considered. Criteria, on the other hand, will often need insights from a specialist to be clearly defined. For example, if the project's overall strategy is to be as sustainable as possible, a factor could be carbon reductions, and the criteria could be: use as little concrete as possible for a specific decision (Christensen, 2022). However, in a different context, the same factor could result in a different criterion, for example, using more low-strength concrete is better. So, specialist insights are often needed to translate the purpose and factors to specific criteria. Decision purposes and factors are mainly defined top-down, and criteria are defined bottom-up. However, it should be done in a process where both perspectives impact the definition of the other.

When the criteria are defined, it should be agreed which are the must-have criteria and what constitutes the want-have criteria. This should also be done in a group conversation. A must-have criterion can discard many alternatives. For example, if the schedule is not to be negatively impacted, all alternatives that might put the schedule at risk should be discarded.

If the vision for the project is clear and tangible, the purpose of the decision is well understood, and if the team is experienced in using CBA as a decision-making method, this process will be easier to manage. The bottom line is that we must explore many alternatives, avoiding pushing for only one alternative. We often start with a much longer list of factors than is needed then clearly state why we choose one alternative over another. We need knowledge and expertise to define the criteria. Additionally, the facilitator role is crucial here to navigate conversations with stakeholders and ensure a transparent process where all voices are included (Arroyo et al., 2014).

Decision-making and documentation

Most of the decision process is completed when alternatives are defined, factors are agreed upon, and criteria are specified. All of the above is preparation for the decision and steps that we often neglect. If you insist on preparing the decision, it will be easy for the team to define the attributes and advantages. Finally, the importance of advantages should be either set by or approved by the decision owner or the paying client.

The recommendation or decision could always be documented in a clear and simple way to ensure transparency. A great tool to document decisions is an A3. A3 is a problem-solving method extensively used in the Toyota Production System, which puts P-D-C-A (Plan, Do, Check, Act) in action. This structured approach helps the problem solvers clearly define the problem, the desired outcomes, the hypothesized solution, and the steps to implement the solution (Shook, 2008). This is done by forcing the problem solvers to summarize the entire issue on an A3 piece of paper (this was the biggest faxable size of paper), which allows for feedback among stakeholders. In this case, decision-makers can use an A3 to present the decision in a clear template (Figure 9.1).

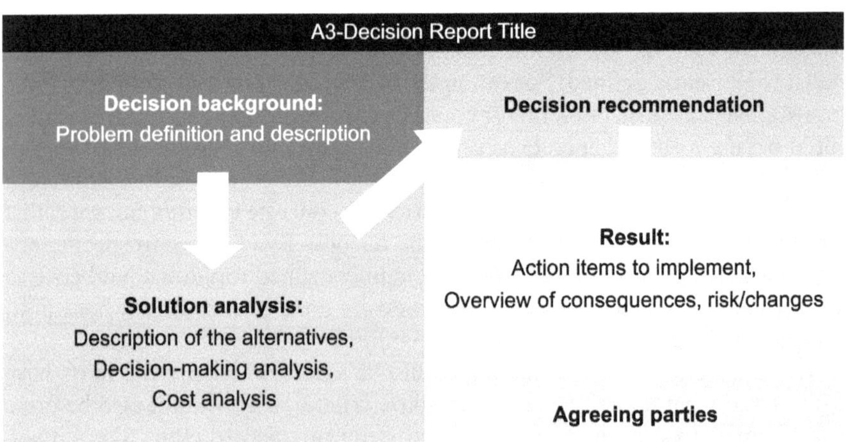

Figure 9.1 The A3 process

Implementation and learning

As we saw in Chapter 1, a decision leads to actions and consequences. While we normally document the one alternative we decide on, we often forget to define what actions we expect to follow this decision. The actions of the decision should be integrated into other systems. It could be that you have an agreed-upon process for introducing changes, including changes to, for example, the schedule or the budget.

If you have prepared the team for collaboration and included relevant stakeholders in the decision, then implementing the decision should be straightforward for all involved, and you could expect the team to support the implementation. The stakeholders and team members might have different opinions and perspectives, but if a proper decision-making process is followed and the documentation is clear, the team will have a consensus on why this alternative was chosen and thus will support the decision. This refers to "The why," as explained in Chapter 2.

It can be helpful to allow for a reconsideration period before moving to action. Here, the central question is: can you improve the decision outcome in any way? If the decision was complicated and had many views, getting time to reflect and consider might bring more ideas for adding value or reducing the cost.

Another aspect to remember is to follow up on the expected outcome and consequences of the decision. As we described in Chapter 2, this is a golden learning opportunity (Christensen, 2008). Did the risk and assumptions play out as expected? Did the decision lead to the expected outcome? This could be done through, for example, the risk or benefits register, if such exists, or by ensuring some lessons learned are planned after a period. A systematic follow-up on actions, risks, and benefits will allow the team to learn and adjust their common approach to decision-making in the future.

Example: Integrating CBA with Lean methods for building design

This example shows a design team that used CBA and A3 for many consecutive decisions in programming, mechanical and geotechnical systems, site safety, and other design elements. The team was able to provide more value to the customer, decrease the initial budget, and make decisions faster over time. Paz was one of the Lean coaches supporting the team in learning the CBA process and the other Lean methods mentioned (Arroyo and Long, 2018). The project was ongoing when they decided to try something new.

The project was a large and complex office building for a technology company in California. The design team was multi-disciplinary and based across multiple countries. Communication was done primarily through video conferencing and regular on-site working periods. The team had minimal or no previous exposure to Lean principles. The project was complicated by an aggressive schedule and budget targets, a difficult regional authority with jurisdiction, and a unique and iconic architecture. The project included the owner, design architects, construction managers, engineers, and trade partners. The owner's representatives were engaged in the decision-making process as the "final decision-makers" or as advisors to higher-level decision-makers based on the A3 with CBA. Not all design decisions were made with the help of the A3 with the CBA process, and the owner did not participate in developing an A3 until the review cycle. The Lean coaches assisted the construction management firm to incorporate Lean tools and methods in the design phase, including Target Value Design, pull planning, and A3 thinking with the CBA decision-making process. The client was particularly concerned with making sound decisions acceptable to multiple, cross-functional stakeholders at all levels within the organization.

Before the intervention, the team faced the primary challenge of a lack of a collaborative decision-making process. Teams were using traditional decision-making techniques that inhibited innovation despite the complexities of the building design. The initial communication patterns, resulted in: (1) hasty decisions based on ungrounded opinions without considering all relevant facts about the performance of design alternatives, (2) misalignment between team members, (3) overturned decisions causing rework, and (4) excessive disconcerted information overwhelming team members. In summary, the lack of structural organization, decision tracking, and methodology for conflict resolution resulted in discernible design inefficiencies.

The coaches helped set a collaborative decision-making process slowly by strategizing with the project manager first, observing the team in action, and then coaching the team one decision at a time. By the end of that design phase, the team had made 24 collaborative decisions using CBA and A3s to manage the discovery process and validate the design team's recommendations to the owner (Table 9.1).

The design budget and value changed along with design decisions and improvements to the baseline design. In most cases, the project cost decreased after a decision as the team was able to find a valuable solution that reduced the baseline cost, but in three instances, the cost of an alternative increased as the team kept selecting the alternative that maximized value for the client. Figure 9.2 shows the variation in project cost with each decision made over time.

Most decisions resulted in cost savings; all A3s documented decisions resulted in approximately $9.7 million in savings, or 11%t of the budgeted amount for these items. Overall, the process resulted in $96,468 in savings per meeting or $12,596 per meeting hour, dividing the savings by the total hours spent in meetings. Regarding time invested for each decision, early CBA-A3 averaged 5.3 meetings per decision and reduced to 3.3 meetings per decision, representing a 37% meeting efficiency increase. The design team members came to the meetings more prepared. They developed a common language with which they more effectively communicated how to make decisions as well as why they were making the decisions.

Table 9.1 List of design decisions documented

Decision title	*Decision title*
1. Modular vs. Non-Modular IDF Closets (An intermediate distribution frame)	13. Soils Management
2. Cistern Single vs. Double-Walled Construction	14. Select the Mix of Fill Material for the Basement Perimeter
3. L2 Zoning Requirements for Open Office and Enclosed Studios	15. Auger Pressure Grouted vs. Precast Concrete Piles
4. Modular vs. Non-Modular Approach for Distribution Electrical Rooms	16. Access to Basement Bike Storage
5. Exhaust Locations for Basement AHUs	17. Security and Maintenance Railing at CUP Opening
6. Underground Infrastructure Support on Suspended Slab	18. PG&E Access to the Main Electrical Room
7. Cistern Sizing Evaluation	19. Location of Outdoor Fitness Area
8. Energy Pile Evaluation	20. Safety Protection at CUP Opening
9. Day 1 vs. Day 2 Lab Loads–Building System Assumptions	21. Return HVAC Zones at L1 and L2
10. Waterproof Membrane Evaluation	22. L1 Zoning Requirements for Conference Rooms
11. Vapor Intrusion Evaluation	23. First Flush vs. Pre-filtration for Canopy Drains
12. Vapor Mitigation Strategies	24. Design of Underground Utilities to Prevent Settlement Displacement

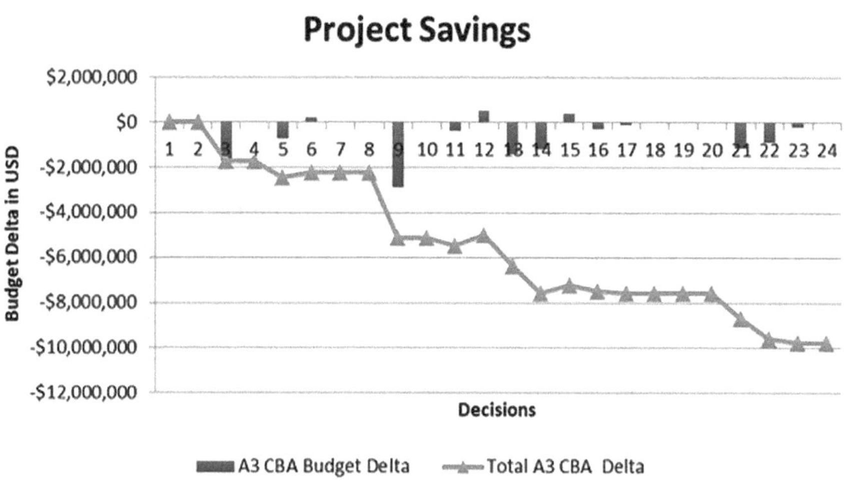

Figure 9.2 Project budget after each collaborative decision
Source: Arroyo and Long (2018).

The design team reported increased client satisfaction, better decision documentation, and more lasting, logical, and reasonable decisions with increased design efficiency and speed. In addition, CBA helped them to have productive conversations, avoid tangential discussions, and focus on differentiating facts from opinions. They developed a shared understanding of the criteria for judgment, and how to compare the advantages of alternatives based on the project context. The relevance and importance of coaching became evident as the decision-making process was contextualized as a time for conversation. Decisions shaped by these conversations came to be seen as commitments to a shared vision of the future. This resulted in an environment where the team learned quickly despite the project manager's limited prior training in guiding a team through this decision-making process. It was important that the design team had flexibility in their decision-making methods and that conversations had space to develop, based on new opportunities rather than being mechanical. We observed a clear need for increased training on decision-making methods at the earliest possible time in the project.

This case illustrates how methods, such as CBA and A3, can enhance collaboration and support a design team in making decisions more efficiently. The study reveals that by utilizing CBA and A3 reports, the design team achieved significant benefits, including increased value for the customer, reduced initial budget, and faster decision-making processes. The study (Arroyo and Long, 2018) emphasizes the importance of collaboration in decision-making and highlights the need for early training in decision-making methods to optimize project outcomes. The team improved decision-making outcomes by setting a collaborative decision process.

Reflection on the decision-making process

We hope our recommendations inspire you to actively engage in decision-making, no matter your role or responsibilities. As an investor or client, you might find inspiration to set up a decision-making process that allows for collaboration and improvements. As a designer, project manager, or contractor, you might find inspiration to plan out more efficient decision-making by engaging the specialists at the right time and presenting the decision in a clear way to senior managers or investors. Finally, as a specialist, you might find inspiration on how best to contribute your knowledge to decisions while accepting that there might be different perspectives to value.

Summary

Defining a clear decision-making process helps teams manage decisions in a way that supports collaboration and reduces iterations. The process also helps manage budgets and schedules comprehensively to create more value for clients.

Key points

- Define a decision and its purpose and include why people should put time and effort into collaborating in this decision.
- Define clear roles and responsibilities in decision-making and set out a transparent governance process.
- Prepare the team to collaborate. You may choose from different approaches to learning CBA. You can have an experienced facilitator or decide to dedicate time to learning so you can keep using CBA with the team.
- Plan for your decisions and when to communicate status and outcome clearly.
- Documenting the decision clearly can help build support for its result. It also helps if things change; you can modify a decision without starting again from scratch.
- Ensure you provide time for reconsideration and create clear actions to implement your decision once it is final.

References

Arroyo, P., Ballard, G., and Tommelein, I.D. (2014). "Choosing By Advantages and Rhetoric in Building Design: Relationship and Potential Synergies." In *Proceedings of the 22nd Annual Conference of the International Group for Lean Construction*, pp. 391–408.

Arroyo, P., Christensen, R., Schöttle, A., and Long, D. (2019). "Lessons Learned on Teaching Choosing by Advantages." In *Proceedings of the 27th Annual Conference of the International Group for Lean Construction (IGLC)*, pp. 405–416. doi.org/10.24928/2019/0249

Arroyo, P., and Long, D. (2018). "Collaborative Design Decisions." In *Proceedings of the 26th Annual Conference of the International Group for Lean Construction*, pp. 463–472.

Ballard, H.G. (2000). "The Last Planner System of Production Control." Doctoral dissertation, University of Birmingham.

Christensen, R. (2008). *Development Practically Speaking: Learning Processes in the Danish Construction Industry*. Department of Production, Aalborg University.

Christensen, R. (2022). "Defining More Sustainable and Innovative Solutions Through Choosing by Advantages." In *Proceedings of the 30th Annual Conference of the International Group for Lean Construction (IGLC)*, pp. 588–598. doi.org/10.24928/2022/0163

Christensen, R., Arroyo, P., and Schöttle, A. (2023). "Perspectives on Roles and Responsibilities of Project Team Members to Enable Collaborative Decision-Making Process." In *Proceedings of the 31st Annual Conference of the International Group for Lean Construction (IGLC31)*, pp. 954–965. doi.org/10.24928/2023/0207

Dreyfus, S.E., and Dreyfus, H.L. (1980). "A Five-Stage Model of the Mental Activities Involved in Directed Skill Acquisition." California University Berkeley Operations Research Center, No. ORC-80-2. Berkeley, CA.

Kahneman, D. (2011). *Thinking, Fast and Slow*. London: Penguin Random House.

Schöttle, A. (2022). "Voraussetzungen und Anreize für die kollaborative Lean-Projektabwicklung im öffentlichen Bausektor: Stärkung der Autonomie, Kompetenz und Verbundenheit von Projektteams." Doctoral dissertation. Karlsruher Institut für Technologie (KIT). 10.5445/IR/1000152023

Schöttle, A., Arroyo, P., and Christensen, R. (2018). "Demonstrating the Value of an Effective Collaborative Decision-Making Process in the Design Phase." In *Proceedings of the 26th Annual Conference of the International. Group for Lean Construction (IGLC)*, pp. 899–909. doi.org/10.24928/2018/0500.

Shook, J. (2008). "Managing to Learn: Using the A3 Management Process to Solve Problems, Gain Agreement, Mentor and Lead." Lean Enterprise Institute.

Suhr, J. (1999). *The Choosing by Advantages Decisionmaking System.* Westport, CT: Quorum.

Ward, A., Liker, J.K., Cristiano, J.J., and Sobek, D.K. (1995). "The Second Toyota Paradox: How Delaying Decisions Can Make Better Cars Faster." *Sloan Management Review*, 36(3), 43.

10 Decisions and innovations for our future

Decision-making can support you in acting responsibly and in accordance with your values. The factors you choose to assess the alternatives and the weight you give to the advantages are tangible indicators of your preferences. In this chapter, we will present why sound decision-making is crucial for ensuring a response to the global crises regarding climate change, biodiversity, and social inequality. We will elaborate on some factors that could be included in decision-making to demonstrate the ability to "walk the talk." We will also show how CBA and the decision-making ideas from this book can be a framework for innovation processes, improving practice in a desired direction and enabling the development of alternative and more sustainable solutions.

A wider perspective

Scientists have defined the boundaries for the planet's ability to regenerate itself and still provide stable conditions for living creatures, including humans, to sustain their existence (Figure 10.1). We should be careful not to contribute further negative impacts to these nine boundaries and instead strive to restore the damage already caused by human beings. A recent update of these boundaries stated that six out of nine have already been transgressed (Richardson et al., 2023). This means that within these areas, we are now outside of the "safe operating space," and if human activities further impact the delicate balance in ecosystems, it could lead to irreversible changes.

Planetary boundaries or other recognized taxonomies, such as the EU taxonomy (European Union, 2020), can inspire the definition of factors used to assess alternatives on a grand scale. Having such a shared global perspective makes a difference for change. For example, since the Montreal Protocol was enacted in 1989, all chemicals that depleted the ozone layer have been phased out (UN, 2023), and we have successfully turned around the harmful impact on the ozone layer. As we now face more major challenges caused by the accumulated harmful human impacts, we should expect more legally binding constraints and thresholds stemming from international agreements, such as the Paris Climate Agreement or the Kunming-Montreal Global Biodiversity

DOI: 10.1201/9781003518310-11

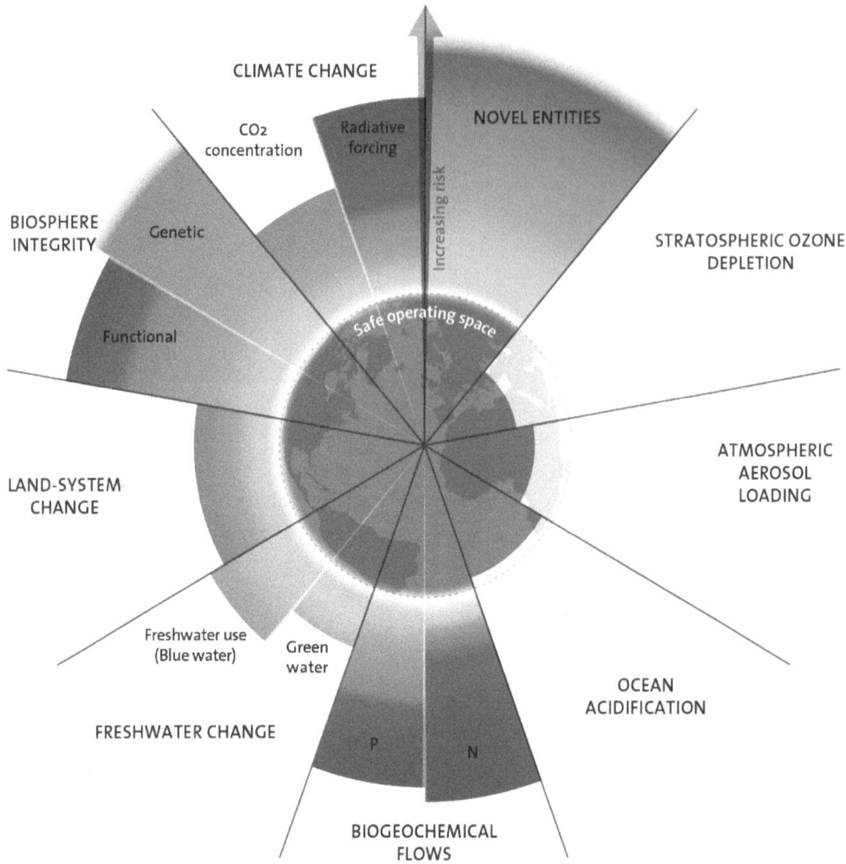

Figure 10.1 Planetary boundaries

Source: Azote for Stockholm Resilience Centre, based on analysis in Richardson et al. (2023).

Framework. For example, climate and environmental considerations must be weighted at least 30% in public procurement in Norway from 2024 (Norwegian Government, 2023), and developers must deliver a biodiversity net gain of 10% in the UK to get planning permission (UK Government, 2023).

Legal requirements might set a direction, but they are still just setting minimum criteria for future projects, and much more can be done to increase their value in the specific context. Projects last for years or even decades, and for some, the change in public opinion is more critical than the legislation at the time of getting the permits. As described in Chapter 8, we can encourage owners to apply Choosing By Advantages (CBA) in tendering. This will create more transparency and help ensure that owners focus on the areas where their project can deliver the most value and impact. It will allow owners to set clear must-have criteria beyond legal requirements. This gives the team a clear understanding of the owner's objectives and expectations. Owners could also clearly point to want-have criteria, allowing the project teams to be creative and collaborate toward more sustainable solutions. In the following section, we will point to a few factors to help inspire a much-needed improvement in the construction industry.

Factors to consider by the construction industry

Where the industry historically has focused on reducing only monetary waste, we now also need to consider how we reduce the negative impact on the environment, people, and climate (Arroyo and Gonzalez, 2016). However, the construction industry extensively impacts our world, including impacts within several planetary boundaries. Here, we will give a small overview of how the industry impacts some planetary boundaries and give ideas on relevant factors to consider if you want to show more sustainable solutions.

The construction industry accounts for 39% of global carbon emissions, where 11% stems from materials, also called embedded carbon (WorldGBC, 2019). Typically, 70–80% of the CO_2 from construction comes from cement, steel, and diesel, so it would matter to set some requirements here. For example, the industry primarily uses cement and steel, and 3% of global CO_2 emissions stem from cement and 7% from iron and steel manufacturing (Ritchie et al., 2020). So evidently, the emissions of carbon can and should be considered when deciding on different alternatives. It is not yet the norm, but more owners are setting specific requirements for carbon reductions (Christensen, 2023). For example, Copenhagen Metro has the ambition to half the carbon footprint of their upcoming new M5 metro line, leading the teams to be creative and develop more sustainable alternatives throughout the project (Smith, 2023).

Another factor to consider is the circularity of materials. The construction industry consumes approximately 50% of global steel production and is the biggest consumer of raw materials. Around 40% of resources extracted globally are materials used for housing, construction, and infrastructure (WorldGBC, 2021), and 40% of the world's solid waste also stems from this industry (Breene, 2016). Cement is the second most used material in the world, next to water. Hence, the construction industry should take responsibility for reducing the use of materials and reusing existing materials, such as sand and gravel, of which a shortage is predicted in many places, and aggregates, therefore, must be transported over long distances due to shortages locally. Setting some requirements for bringing forward alternatives to promote a circular economy could be advantageous for the owner, as this requires the value chain to collaborate.

Protecting and enhancing biodiversity are also a challenge for the industry. The construction industry is globally responsible for nearly 30% of biodiversity loss (WEC, 2020). The impact happens in several phases, from the exploitation of raw materials to construction, where a site might be damaged, and operation and permanent land use change. Constructing new infrastructure projects such as roads, harbors, and tunnels has likely destroyed existing habitats and ecosystems. However, much can be done by protecting and improving habitats to decrease the negative impact or enhance biodiversity, as governments will increasingly demand. Enhancing biodiversity in an area could also be a natural way to mitigate climate adaptation risks, such as flooding and landslides.

The industry also has an extensive impact on human lives for those working in the industry and those impacted by the end products. The construction industry holds an unwanted record of being the industry with the most accidents and fatalities. Worldwide, construction workers are three times more likely to be killed and twice as likely to be injured than workers in other occupations (Phillips, 2023). It is the industry with the second highest numbers of suicides in the US (Sussel et al., 2023). For these reasons, we should avoid the hard price competition, where the lowest tender usually wins, and recognize the impacts on the last person in line. Health, safety, and well-being have been a focus in the construction industry for many years. However, we still fail to make systematic and lasting improvements.

Despite the challenges, the industry is also providing many benefits and is needed to help equal development. Globally, we need more new housing and infrastructure structures than ever before. More than half of the global population lives in cities; by 2050, this is expected to increase to 70%. This will lead to 230,000 km^2 of new buildings being built, and it is estimated that 75% of the infrastructure needed in 2050 is yet to be built (WorldGBC, 2021). We predict that the construction industry will continue to grow massively to deliver the needed housing and infrastructure for urbanization and enhance the livelihood of many, particularly in the Global South.

As part of transitioning from fossil fuels to a more sustainable alternative, we need extensive infrastructure to support electrification. It has been predicted that we need $139 trillion invested in green infrastructure to reach net zero by 2050 (FIDIC, 2023). The materials needed to build all this will increase immediate CO_2 emissions, so responding to the need for equal development and green transition is not without dilemmas. On one side, the industry needs to improve livelihoods and support the transition to more sustainable energy production, but in doing so, we also need to address the massive negative impacts of the industry.

The data above will quickly change, and experts may have differing views on details. Yet, each project will have its specific context and areas where careful consideration can add value. The significant negative impact of the industry calls for innovative solutions. We must consider how we can increase circularity, reduce waste, and impact the climate and environment while creating livable societies for humans. There are no straightforward answers here. You can improve on one factor while failing on another, and this complexity of competing challenges is difficult for many to navigate. Not all factors are numeric, and the assessment of value will often be subjective. It can be difficult and scary to step into this complexity, where you realize that known solutions are no longer sufficient, but new solutions also come with a trade-off. However, with an improved focus on decision-making and the right skills, we can allow for more transparency and systematically learn and develop, as shown in the next sections.

Sustainable decision-making for change

There can be several reasons for going beyond legal sustainability requirements. You might want to take responsibility for driving the industry more sustainably and lead by example. Another reason could be to ensure a healthy investment, as more financial institutes are giving advantageous loans and returns linked to sustainability parameters. A third reason could be a wish to brand your project or organization and back this with signing up to a pledge, ensuring integrated annual reporting, or certifications. Finally, within the construction industry, being more sustainable is often equal to using fewer materials and less land; thereby, it becomes a good financial business case that ensures more cross-disciplinary collaboration.

No matter the reason, if you want to claim to deliver beyond what could be expected, you must clearly document the change you have made, or you will risk being accused of greenwashing. Here, decision-making can be key. A transparent decision-making process will ease your need to document that you do more than what could be expected. By choosing between alternatives, including a baseline based on normal practice, and by assessing the alternatives based on sustainability factors as described above, you can demonstrate that you have chosen a more sustainable solution.

Companies or public entities wishing to brand themselves as supportive of more sustainable business can sign up for different accreditations and pledges. With such promises and pledges, organizations demonstrate a will to go beyond legislative requirements and document their achievements with third-party verification and certification. Such pledges and certifications create a shared understanding of how to manage sustainability. Documenting decisions often counts positively for those certifications, helping drive development through real impact and changes. However, for some, certification could end as a check-the-box exercise where getting the right level is the main goal, but if the certification process is supported by healthy decision-making, it could be an efficient way to get credit for certain achievements.

We should recognize that major decisions should be considered carefully, such as whether to invest in new infrastructure or renovate the existing one. However, a sound decision-making process is also important for the thousands of minor decisions within projects, as they have an enormous cumulative impact, such as what material to use, what methods to apply, and whether to reduce the size of a structure. Decisions at all levels matter; we cannot afford to make unsound decisions anymore, as our window of opportunity is closing.

Luckily, we know most of the technologies and changes needed, and we "just" need to scale these and ensure implementation. We believe decision-making is key to accelerating this development.

CBA to identify alternatives and drive innovation

A decision-making method alone won't make more sustainable decisions. However, it could help argue the case for more sustainable alternatives. The previous section showed the value of using factors and criteria to choose more sustainable alternatives. To be able to do so, we will also have to question whether we are choosing between the right alternatives or if we should look for or invent more alternatives. With the pressure many feel in the industry, we could move forward with a valid solution without looking for alternatives. However, new alternatives need to be created, identified, and considered to instigate real change.

The alternative we usually do not consider is perhaps the most obvious: Build nothing or do nothing. Doing nothing has the least impact on material consumption, carbon emissions, and the use of raw materials. Do we need another new structure? Or could our needs be covered in alternative ways, such as by sharing a building or changing local transport patterns to reduce rush hour flows? Do we have another reasonable solution at hand? Modernizing an existing bridge instead of tearing it down and rebuilding it might still need materials, but far less. Using alternative or reused materials could also reduce the need for virgin materials and carbon emissions.

Systemically considering alternatives at critical points in a project could slow down your thinking as a group. In Norway, for example, the concept phase of a publicly financed megaproject requires an assessment of at least three different alternatives to get funding, including two conceptual alternatives along with a minimum solution, including just enough to get through (Finansdepartementet, 2023). For example, if you are seeking funding for a road crossing over a river, you might present two types of bridges and a "do nothing" solution where you reduce the peak commuting demand. Comparing these alternatives allows everyone to consider the arguments for the preferred solution carefully, and it might even cause the team to change their opinion.

Alternatives for more sustainable solutions should be considered at all levels. However, recognizing or creating alternatives is not easy. We are used to focusing on solving problems using known technologies, and as time is often limited, we are often satisfied with a viable solution. Giorgio Parisi (2021), the Nobel Prize-winner in physics, says that micro creativity, meaning small, everyday ideas and innovations, is fundamental for making progress. New ideas and concepts do not appear out of the blue by magic; they require the right setting, a small idea to spark creativity, hard work, and a bit of luck. Here, creativity plays an important role, and, often, ideas stem from the unconscious, where we make connections between different contexts. CBA could be used to recognize and create new alternatives by setting out the playing field and enabling people to bring ideas from other contexts. The following example illustrates this idea.

Creating innovative alternatives with CBA

In this example, an owner invited industry partners to develop conceptual alternative designs for a standard metro station. The owner's aim was to explore new technologies and ideas to reduce the carbon footprint of a standard metro station compared to a baseline project. The result of this desk study, an early investigation to identify alternatives, was later used as input to design requirements for a new metro line. The winning team of this desk study included a range of specialists: experienced senior engineers and specialists in underground structures, geotechnics, concrete, mechanical, electrical, sustainability, and tunnel design.

Initially, the team had difficulties collaborating on creating new and better alternatives. Every discipline perceived the project from its own perspective and focused on improving the design without knowing how these changes impacted the project more broadly. Some team members had previously used CBA for decision-making, and the idea came about to use the same method to create alternatives based on factors and criteria (Christensen, 2022).

Randi worked as a CBA facilitator, and the team onboarded her. They focused on creating alternatives to meet the success criteria set by the owner and agreed to a modified CBA process to drive innovation and collaboration, using the following steps:

1. Define a baseline: A comparable existing project, including background reports.
2. Define factors: Areas where the owner expected improvements compared to the baseline.
3. Define criteria: Must-have and want-to-have criteria for the alternatives.
4. Define the attributes: Create the baseline using historical data.
5. Create alternatives based on criteria: Ideas from all disciplines were assessed and combined into different alternative solutions.
6. Assess all alternatives using the CBA tabular method, including all relevant factors.

From the brief, it was clear that the owner was looking for alternatives that reduced the carbon footprint. The team, therefore, considered reducing CO_2 emissions throughout the project, including changes in the structural elements, quantities, transportation, materials, and construction methods while keeping the architectural functionality, look-and-feel, and finishes unchanged.

By using this approach, the specialists could contribute with their unique knowledge. The team set the corner flags for innovation using the factors, criteria, and baseline project as the starting point. At this moment, the specialists could give input to changes, bring relevant knowledge and experiences from other international or local projects, and suggest how these could be implemented in these specific contexts. The team integrated ideas into different alternatives by having a shared decision-making process. All disciplines could contribute to comparing the advantages of each alternative. This process also helped some disciplines to be more courageous and bring in bold ideas that would normally be discarded. One example was that the team considered removing structural elements, such as the tunnel's inner lining. This is typically added to improve the structure's watertightness, and many specialists dismissed the idea of removing this, as it was not standard practice and they perceived it as an unnecessary risk to electrical installations. However, others challenged the idea and asked if the lower carbon footprint wasn't more important than this risk, which could be mitigated by extra monitoring with new technology.

The owner's clear vision drove the creative process and allowed the team to collaborate to create new alternatives. One potential pitfall identified in using CBA to create new alternatives was the risk of focusing solely on improvements and forgetting to include factors where the solutions or suggested changes might lead to risks or less value. An experienced facilitator will include more factors or alternatives to frame the different alternatives. In the final assessment, the team included more factors, such as durability, reliability, and robustness, as the reduction in concrete could lower design lifetime, more interruptions of operations, and more maintenance. Other factors considered were construction time and the risk of using new technologies.

Using CBA in this example allowed the team to focus on their discipline while understanding the bigger picture. It gave them a structured process to collaborate and to ensure progress instead of endless iterations. Based on agreed factors, each discipline could suggest specific criteria and develop ideas for improvements and new technologies that pointed in a shared direction. The owner got a summary report on one page (A3) and a more extensive report with a clear and structured presentation of the alternatives, the agreed-upon factors, criteria, attributes, and a comparison. This formed the basis for the owner to decide on the risk profile and set design requirements for the upcoming project.

Leadership competencies for the future

As we have seen above, decision-making as a discipline is a cornerstone for changing our society and creating more sustainable projects. We must consider alternative solutions on all levels and challenge current practices. This also calls for a redefined set of leadership skills. UNESCO has pointed to leadership competencies needed to fulfill the United Nations Sustainability Development Goals (Leicht et al., 2018). In summary, these key sustainability competencies are:

1. Systems Thinking: Ability to analyze complex systems and deal with uncertainty
2. Anticipatory Ability: Being able to understand and evaluate multiple futures
3. Normative Ability: Ability to reflect on norms and values

4. Strategic Action: Collectively develop and implement innovative actions
5. (Emphatic) Collaboration: Learn from and respect others
6. Critical Thinking: Ability to question norms and practices
7. Self-awareness: Reflect on own role and impact
8. Integrated Problem-solving: Ability to apply different problem-solving frameworks

Through our extensive experience with CBA, we can recognize the many skill sets used in sound decision-making (Table 10.1).

Table 10.1 Overview of how CBA supports sustainable competencies

Sustainability competencies	*How CBA supports this competency*
Systems thinking	CBA methods support system thinking by enabling assessment against multiple factors, which may be interdependent without being directly comparable. Predicting the future based on the information available and risk management is integral to our approach to decision-making.
Anticipatory	As explained in Chapter 1, decision-making involves predicting the future consequences of each alternative. By investigating the attributes of each alternative, we are trying to anticipate the consequences.
Normative	The book, particularly Chapter 2, gives ideas for reflecting on norms, values, and biases. Also, by enabling deeper reflection on decisions, the team can challenge current practices and bring in new ideas. If we value sustainability, we will look for social, environmental, and economic factors to support our decisions. Our values will also drive the alternatives we develop, as presented above. Therefore, our values will be made as explicit as possible, and this strategy will help us check our own biases and assumptions.
Strategic	Decision-making can help mature a vision on a strategic level by questioning the weight of advantages in strategic decisions. A clear vision with transparent objectives will cascade into decisions on all levels and support a creative and innovative culture. Using CBA, decision-makers can develop a strong consensus on the decision outcome, leading to more efficient implementation.
Collaboration	CBA requires the different disciplines to contribute with knowledge and data while also understanding the full picture. Stakeholders will define factors for evaluation; designers and engineers will develop feasible alternatives; experts will need to help specify criteria and present attributes for each alternative; managers and leaders will need to weight the importance of advantages, and financially savvy people will need to help assess initial and long-term costs. By understanding and respecting different inputs, we enable true collaboration and the development of new ideas by building on the collective competencies and ideas of the team.

(*Continued*)

Table 10.1 (Continued)

Sustainability competencies	*How CBA supports this competency*
Critical thinking	CBA allows us to question business as usual and management, question our preferences and values, and challenge our in-built aversion to change. We need to think critically about the context of the decision. A decision cannot be copied and pasted from one country or city to another or from one use to another without critically assessing local and present norms and values by involving local stakeholders to understand their needs and the alternatives they have available and critically assessing norms and practices.
Self-awareness	We must be aware of our decisions and how we make them, which will impact the outcomes. Awareness of how to make better decisions also gives us the responsibility to act. CBA helps separate facts from opinions and value from cost, enabling constructive discussion and reducing conflicts in the team.
Integrated problem-solving	CBA and a collaborative decision-making process efficiently integrate the team's collective knowledge and experience to solve some of the project's specific problems while considering the sustainability challenges.

We see sound decision-making, particularly CBA, as an enabler for transforming our industry toward creating a more sustainable and liveable future. The CBA system will not, on its own, bring more sustainable solutions. However, combined with a strong vision and qualified facilitation, it can be a driver for a more collaborative and innovative culture, and decision-by-decision, we can make the industry more sustainable. As we have seen in this chapter, we have a responsibility beyond the immediate impact of our projects. CBA can practically support making more sustainable decisions.

Key points

- Consider which factors are being used when assessing alternatives. By adding factors that point to long-term social and environmental impacts, we can demonstrate how we include sustainability considerations in our projects. Multicriteria decision-making is a way to assess complex systems while being transparent about the uncertainties included.
- Consider how to systematically assess alternatives. At an early planning stage, we strongly recommend always including the alternative "do nothing" and considering the alternatives at all stages. The accumulated impact of many small decisions is also vital to the change needed.

- Consider how to work systematically with innovation and improvements. If you do not integrate such considerations systematically, we will likely choose "a viable solution" and continue business as usual.

References

Arroyo, P., and Gonzalez, V. (2016). "Rethinking Waste Definition to Account for Environmental and Social Impacts." In *Proceedings of the 24th Annual Conference of the International Group for Lean Construction*, pp. 13–22.

Breene, K. (2016). "Can the Circular Economy Transform the World's Number One Consumer of Raw Materials?" World Economic Forum. Available at: www.weforum.org/stories/2016/05/can-the-circular... (accessed January 14, 2024).

Christensen, R. (2022). "Defining More Sustainable and Innovative Solutions Through Choosing By Advantages." In *Proceedings of the 30th Annual Conference of the International Group for Lean Construction (IGLC30)*, pp. 588–598. doi.org/10.24928/2022/0163

Christensen, R. (2023). "Different Strategies for Setting Requirements for Sustainability." Paper presented at IABSE Congress 2023, Engineering for Sustainable Development, New Delhi.

European Union. (2020). "EU Taxonomy for Sustainable Activities." Available at: www.easa.europa.eu/en/domains/environment/policy (accessed June 2023).

FIDIC (International Federation of Consultant Engineering). (2023). "Closing the Sustainable Infrastructure Gap to Hit Net Zero." Available at: issuu.com/fidic/docs/closing_the_sustainable...

Finansdepartementet. (2023). "Hva er statens projektmodell?" [What Is the State Project Model?]. Available at: Hva er statens prosjektmodell? - regjeringen.no (accessed April 13, 2024).

Leicht, A., Heiss, J., and Byun W.J. (2018). "Issues and Trends in Education for Sustainable Development." UNESCO. Available at: unesdoc.unesco.org/ark:/48223/pf0000261445

Norwegian Government. (2023). "Nå skal klima og miljø vektes minst 30 % i offentlige anskaffelser" [Climate and Environment Must Be Weighted Min 30% in Public Procurement]. Available at: Nå skal klima og miljø vektes minst 30 % i offentlige anskaffelser - regjeringen.no (accessed January 14, 2024).

Parisi, G. (2021). *I en Stæreflok, Sønheden i komplekse systemer* [In a Flight of Starlings, the Wonder of Complex Systems]. Copenhagen: Gutkind.

Phillips, Z. (2023). "Construction Had the Most Fatalities of Any Industry Last Year." Construction Dive. Available at: www.constructiondive.com/news/construction.. (accessed January 14, 2024).

Richardson, K., Steffen, W., Lucht, W. et al. (2023). "Earth Beyond Six of Nine Planetary Boundaries." *Science Advances* 9(37). https://www.science.org/doi/10.1126/sciadv.adh2458

Ritchie, H., Rosado, P., and Roser, M. (2020). "Emissions by Sector: Where Do Greenhouse Gases Come From?" Available at: https://ourworldindata.org/emissions-by-sector [Online Resource].

Smith, C. (2023). "COWI and Arup Challenged to Halve Carbon Footprint on Copenhagen's Planned M5 Metro Line." *New Civil Engineer*, June 13. www.newcivilengineer.com/latest/cowi-and-arup (accessed October 2023).

Sussel, A., Peterson, C., Li, J., Minino, A., Scott, K.A., and Stone, D.M. (2023) “Suicide Rates by Industry and Occupation — National Vital Statistics System, United States, 2021.” December 15, 2023. 72(50), 1346–1350, Centers for Disease Control and PRevention. Available at: www.cdc.gov/mmwr/volumes/72/wr/mm7250a2.htm (accessed April 13, 2024).

UK Government. (2023). “Understanding Biodiversity Net Gain.” Available at: www.gov.uk (accessed January 14, 2024).

UN (2023). “Ozone Layer Recovery Is -n Track, Helping Avoid Global Warming by 0.5 Degrees Celcius.” Available at: unep.org (accessed June 6, 2024).

WEC (World Economic Forum). (2020). “The Future of Nature and Business, New Nature Economy Report II.” Available at: www.weforum.org/publications/new-nature-economy...

WorldGBC (World Green Building Council). (2019). “Bringing Embodied Carbon Up Front.” 404. Available at: worldgbc.org (accessed February 16, 2021).

WorldGBC (World Green Building Council). (2021). “Beyond Buildings Report 2021.” Available at: worldgbc.s3.eu-west-2.amazonaws.com (accessed January 14, 2024).

11 How to build better decisions?

After reading this book, we hope you realize you have many choices and decisions to make. Do not wait passively for others to make decisions; there is too much at stake, especially if you work in the construction industry and realize our decisions' social, environmental, and economic impact. Decision-making is a process, and by now, you know how to engage actively, ask questions, and support creating a fair and inclusive process for better decision-making. Decision-making methods and processes matter; you now know what to look out for in a process and why we recommend Choosing By Advantages (CBA) as a decision-making method.

We would like to wrap up this book by providing advice on how to get started and continue on your path to building decisions that provide a better chance of success and improvement in the construction industry. All change is difficult, but we have been learning, teaching, coaching, facilitating, and implementing CBA for years and have a few tips and tricks for you.

Setting a collaborative decision-making process

No matter what your role is, an investor, a client, a project manager, a designer, a specialist, or something completely different, you have the ability to continuously improve decision-making around you. We are all responsible for nudging, requiring, or requesting a better decision-making process. To support this, we have throughout the book touched on different key elements to remember to support a truly collaborative decision-making process:

- *Be inclusive and listen*: Make sure you include subject matter experts; if you cannot involve all, keep other stakeholders informed. Include the client, users, and people who will maintain what you are building. You can set up the process so people can efficiently contribute with their ideas on alternatives, factors, and attributes to make the decision.
- *Be diverse*: Look around the room where decisions are made and ask: Is this group diverse? Do people here come from the same discipline and have similar backgrounds, genders, ethnicities, ages, and abilities? A more

DOI: 10.1201/9781003518310-12

diverse group will make less biased decisions, spot more risks and opportunities, and be more innovative.

- *Make decisions at the right level*: The more you involve those closer to the problem and those who understand the consequences of the decisions, the better the outcome for the project or organization. Refrain from overloading management with decisions where their input is not needed or where others better understand the decision.
- *Future-proof your decision*: Consider future constraints and factors that future generations would care about. They won't be in the room, but think about your family, kids, nieces, nephews, etc. What flexibility would they appreciate? How do we leave a better planet behind?
- *Be collaborative*: Collaboration in this context means participating in the decision, engaging, and contributing. Leaders have a key role here so everyone can feel psychologically safe sharing ideas and concerns.
- *Be Lean*: Many great Lean management ideas can support making your decision-making process efficient and effective. Plan your decisions logically using a pull system to ensure alignment of expectations on resources to make the decision and to ensure communication of the outcome. Use set-based design to explore alternatives until the last responsible moment. Allow money to cross contract boundaries so you can optimize the whole, not just the pieces. Document your decisions using A3 reports and CBA.
- *Think about new alternatives*: Keep thinking about new alternatives. We will never improve if we keep choosing well-known and consolidated solutions. We need alternatives that take us forward, and we need to be transparent about what we would like to improve.

Benefits of using CBA

We hope that this book has helped you understand why decision-making methods matter. We also hope these ideas have left as big an impression on you as they did on us. For more than 15 years, we have implemented and used CBA methods. We have trained people on how to implement CBA and chose the most suitable CBA method for their context and situation. We also have documented the importance of having a transparent and collaborative decision-making process.

When you want to introduce CBA to your colleagues, boss, or client, you must be clear about what the project team gains by introducing the method. CBA has several benefits, such as:

- *CBA helps focus on value*: When understanding value, you need a collaborative multidisciplinary team. CBA can include and integrate multiple value perspectives in the decision. The team can include multiple factors and transparent criteria for judgment. CBA also can help assess whether the team's decision delivers value to the client.

- *CBA helps to create transparency and buy-in*: When a team uses CBA, they create transparency when choosing an alternative. The team can understand the rationale for a decision. Everyone can see the advantages of each alternative and the trade-offs that decision-makers made. The team can see the value vs. the cost of the alternatives. They can choose the best value for their resources and optimize decisions across the project. Additionally, when people are aware of the rationale behind a decision, they are more likely to act according to it.
- *CBA helps to build consensus*: When teams use CBA methods, they agree on relevant facts based on a decision context, not on abstract factors or ideas, thereby avoiding conflicts and unnecessary iterations.
- *CBA supports continuous learning*: When decisions are clearly documented, it is easy to assess whether the outcome aligns with expectations. This allows learning from the decision-making process. The team can update a decision if new information is available, such as a new factor to consider or an additional alternative.
- *CBA supports innovation*: When evaluating multiple alternatives, practitioners understand the advantages of each alternative. With this understanding, new innovative alternatives can be created, often as a hybrid of previous alternatives that combine more advantages and reduce costs. When the team knows the client's values more clearly, they can better optimize and find new alternatives, creating more value or saving resources.
- *CBA supports sustainable decisions*: When a team or a client values environmental and social aspects, CBA can help them clarify how a more sustainable alternative is better. Decision-makers can establish environmental and social criteria in addition to economic aspects. Separating the cost from value and also further considering initial and lifecycle costs often provides a team with arguments to choose a more sustainable alternative, even if it is not the least expensive in the short term.

Your next decision matters

Knowledge is power. The pain of knowing how to make better decisions collaboratively is that you are now responsible for acting on it.

If you are an owner, an investor in the construction industry, or a politician making strategic decisions for your community or country, you can set requirements for the team to provide you with clear and transparent information to decide; you can ask your team to use CBA. You also know how important it is to communicate the purpose, value, and strategic ambitions for the team to make an efficient decision. You know that cost is a trade-off and that you need to pay as much attention, if not more, to the value creation, to be sure you get more value for the money you have available. You can help the team by actively collaborating to clarify your values and priorities and establish clear must-have and want-to-have criteria. You also know that sometimes, if you want to go fast, you must go slow. When things get complicated and decisions have a huge impact, you have to plan the decisions out and get the right people involved at the beginning. This will lead to a faster and clearer decision process than if you rush the team.

If you are a professional in the industry, you now know how to make your expertise and experience count in decision-making. You need to support a sound decision, an efficient decision process, and a culture where everyone feels safe to speak up. By proposing the right alternatives, making sure the factors and criteria point to what matters, and providing accurate attributes to highlight the advantages of each alternative, you can help the team and the investors make timely and efficient decisions. You also know that everyone does not have to agree, but everyone should feel included and understand the outcome of the decision to enable implementation of the outcome.

We hope this book has equipped you to make a difference if you are new to the industry. Our industry needs bright minds from various backgrounds with different skill sets and experiences. We need you to play an active role from day one. We hope this book gives you the courage to initiate change constructively by participating in and one day leading decisions that value the diverse construction industry's experiences and skills to create a more sustainable future.

Final remarks

Regardless of your role, or whether or not you work in the construction industry, you will be affected by the decisions people make while designing and building. Next time you use a highway or a building or even while being in your house, you may wonder why it was built this way. We build decisions, literally and figuratively. We would like to bring awareness to everyone that the way we set the decision-making process and the decision-making methods we use matter. Both the decision-making process and method will impact the outcome of your decisions.

The time is right for a book that provides theory, research, and practical implementation on collaborative decision-making. Fortunately, more owners are requesting a more collaborative approach to building construction projects, which requires a collaborative decision-making process. It is time we start questioning our decision-making practices and understand that decision-making is not solely a matter of authority. We need to enable people to collaborate to reach the best possible outcome and have an efficient implementation.

We, Paz, Annett, and Randi, will continue our journey in our respective jobs in different countries and continents. As three senior engineers and curious researchers in the construction industry, we will continue to apply sound decision-making and learn, teach, and coach more in doing the same. We will continue making keynote speeches at conferences worldwide and communicating with senior managers in companies, at the board level, and with politicians to focus on how we make decisions because it matters.

We would like to invite you to join this journey, and if you want to learn more about making better decisions, you can follow us at CollabDecisions.com. Our mission is to inspire and support the construction industry in making better and more collaborative decisions. Our vision is a diverse and inclusive construction industry that embraces collaboration in the decision-making process to innovate and improve the industry's impact on society now and for future generations.

We encourage you to build on our experience and research and share your experiences with us and the world. Together, we will make countless better, collaborative, transparent, and important decisions.

Glossary

A3 A problem-solving method extensively used in the Toyota Production System, which puts P-D-C-A (Plan, Do, Check, Act) in action.

Advantage A benefit, gain, improvement, or betterment. The beneficial difference between the attributes of two alternatives (one of which is the least preferred).

Anchor-Statement Advantage The difference between the minimum requirement attribute and the best expected attribute.

Alternatives Two or more people, things, or plans from which one is to be chosen.1

Attribute A characteristic, quality, or consequence of one alternative.

Best expected attribute The best attribute a proposal is expected to have on a given factor in the tendering process.

Bias Inclination or prejudice for or against one person or group, especially in a way considered unfair. Several types of biases exist, as described by the SEEDS Bias Model.

CBA (Choosing By Advantages) A decision-making system based on the importance of advantages, developed by Jim Suhr (1999).

Cognitive dissonance Occurs when there is an uncomfortable tension between two or more beliefs held simultaneously. This most commonly occurs when our action and behavior do not align with each other—we believe one thing but act against those beliefs.

Criterion A standard on which a judgment is based. It can be a *must-have* criterion (i.e., an alternative must comply with maximum or minimum criteria; otherwise, it is discarded) or a *want-to-have* criterion (e.g., more or less is better).

Decision-makers The ultimate party responsible for deciding and defining the actions required to move forward.

Factor An element, part or component of a decision.

Importance of Advantage (IoA) An assessment of the importance of the advantage, given on an agreed-upon numerical scale. Typically used from 0 to 100.

Integrated Project Delivery (IPD) A collaborative construction project delivery system in which key parties involved in the design, fabrication, and construction aspects of a project are joined together under a single agreement. IPD involves the knowledge of experts in the project as early as possible and aligns stakeholders' incentives through shared risk and reward.

Last Planner System (LPS) The Last Planner System™ (Ballard 2000) is a well-known lean construction management practice. The basic principle is to involve the people doing the work (the last planners) in planning their own work based on pull mechanisms.

Last responsible moment The moment where postponing a decision further will cause irreversible consequences, but doing it earlier will be based on non-updated information.

Lean Management philosophy focusing on value and continuous improvement.

Lifecycle cost All costs related to an investment, including initial investment, maintenance, operation, dismantling, and transport of materials to be reused.

Minimum requirement attribute The minimum attribute a proposal is required to have on a given factor. Otherwise, the proposal will be discarded from the tendering process.

Net Zero Energy (NZE) building A building that produces the same energy that it uses.

Paramount Advantage The most important advantage.

Prior Anchoring Process (PAP) Is an add-on to the CBA Tabular method for the public tendering procedure published in the tendering document as part of the request for proposal so bidders understand how their proposal will be evaluated.

Psychological safety A belief that one will not be punished or humiliated for speaking up with ideas, questions, concerns, or mistakes, and that the team is safe for interpersonal risk-taking (Edmondson, 2018).

SEEDS Bias Model Classification of biases in similarity, expedience, experience, distance, and safety bias.

System 1 thinking Thinking fast and automatically, making decisions unconsciously and with little effort (Kahneman, 2011).

System 2 thinking Thinking with attention and effort, making decisions consciously requires mental effort (Kahneman, 2011).

Takt The constant frequency and amplitude of a beat. Takt Planning is a visual scheduling method that divides the project into zones and allocates time-based tasks, helping teams see flow and be more efficient.

Tender A process by which the client decides what team to work with based on different factors, such as experience, references, quality, and price.

Value Specific to the worth of what a thing or a person provides or brings to a situation, and it is judged from the client's perspective.

Values Central beliefs and ideals that guide you in your life, for example, honesty, family, hard work, inclusion, integrity, kindness, impact, etc.

References

Ballard, G. (2000). "Positive Vs. Negative Iteration in Design." Paper presented at 8th Annual Conference of the International Group for Lean Construction. Brighton, UK, 17–19 July.

Edmondson, A.C. (2018). *The Fearless Organization: Creating Psychological Safety in the Workplace for Learning, Innovation, and Growth*. Hoboken, NJ: John Wiley & Sons.

Kahneman, D. (2011). *Thinking, Fast and Slow*. London: Penguin Random House.

Suhr, J. (1999). *The Choosing By Advantages Decisionmaking System*. Westport, CT: Quorum.

Index

Note: Page numbers in *italic* type refer to figures; those in **bold** type refer to tables.

FSC
www.fsc.org
MIX
Papier | Fördert
gute Waldnutzung
FSC® C083411

Kolibri 360 GmbH
Ferdinand-Jühlke-Straße 7
99095 Erfurt, Deutschland
produktsicherheit@kolibri360.de